COMBINATORIAL MATHEMATICS

By

HERBERT JOHN RYSER

THE
CARUS MATHEMATICAL MONOGRAPHS

Published by
THE MATHEMATICAL ASSOCIATION OF AMERICA

Committee on Publications
E. F. BECKENBACH, Chairman

Subcommittee on Carus Monographs
D. T. FINKBEINER II, Chairman
R. P. BOAS
ARGELIA V. ESQUIVEL
GEORGE PIRANIAN

THE CARUS MATHEMATICAL MONOGRAPHS are an expression of the desire of Mrs. Mary Hegeler Carus, and of her son, Dr. Edward H. Carus, to contribute to the dissemination of mathematical knowledge by making accessible at nominal cost a series of expository presentations of the best thoughts and keenest researches in pure and applied mathematics. The publication of the first four of these monographs was made possible by a notable gift to the Mathematical Association of America by Mrs. Carus as sole trustee of the Edward C. Hegeler Trust Fund. The sales from these have resulted in the Carus Monograph Fund, and the Mathematical Association has used this as a revolving book fund to publish the succeeding monographs.

The expositions of mathematical subjects which the monographs contain are set forth in a manner comprehensible not only to teachers and students specializing in mathematics, but also to scientific workers in other fields, and especially to the wide circle of thoughtful people who, having a moderate acquaintance with elementary mathematics, wish to extend their knowledge without prolonged and critical study of the mathematical journals and treatises. The scope of this series includes also historical and biographical monographs.

The following monographs have been published:

No. 1   Calculus of Variations, G.A. BLISS

No. 2. Analytic Functions of a Complex Variable, by
        D. R. CURTISS

No. 3. Mathematical Statistics, by H. L. REITZ

No. 4. Projective Geometry, by J. W. YOUNG

No. 5. A History of Mathematics in America before 1900, by
        D. E. SMITH and JEKUTHIEL GINSBURG
        (out of print)

No. 6. Fourier Series and Orthogonal Polynominals, by
        DUNHAM JACKSON

No. 7. Vectors and Matrices, by C. C. MacDUFFEE

No. 8. Rings and Ideals, by N. H. McCOY

No. 9.   The Theory of Algebraic Numbers, Second Edition, by
         HARRY POLLARD and HAROLD G. DIAMOND

No. 10. The Arithmetic Theory of Quadratic Forms, by
         B. W. JONES

No. 11. Irrational Numbers, by IVAN NIVEN

No. 12. Statistical Independence in Probability, Analysis
         and Number Theory, by MARK KAC

No. 13. A Primer of Real Functions, Third edition, by
         RALPH P. BOAS, Jr.

No. 14. Combinatorial Mathematics, by HERBERT JOHN
         RYSER

No. 15. Noncommutative Rings, by I. N. HERSTEIN

No. 16. Dedekind Sums, by HANS RADEMACHER and
         EMIL GROSSWALD

No. 17. The Schwarz Function and its Applications, by
         PHILIP J. DAVIS

No. 18. Celestial Mechanics, by HARRY POLLARD

No. 19. Field Theory and its Classical Problems, by
         CHARLES ROBERT HADLOCK

No. 20. The Generalized Riemann Integral, by
         ROBERT M. McLEOD

No. 21. From Error-Correcting Codes through Sphere
         Packings to Simple Groups, by
         THOMAS M. THOMPSON

No. 22. Random Walks and Electric Networks, by
         PETER G. DOYLE and J. LAURIE SNELL

No. 23. Complex Analysis: The Geometric Viewpoint, by
         STEVEN G. KRANTZ

The Carus Mathematical Monographs

NUMBER FOURTEEN

COMBINATORIAL MATHEMATICS

By

HERBERT JOHN RYSER

Professor of Mathematics
Syracuse University

Published and Distributed by
THE MATHEMATICAL ASSOCIATION OF AMERICA

© 1963 by
The Mathematical Association of America

Complete Set ISBN 0-88385-000-1
Vol. 14 ISBN 0-88385-014-1

Library of Congress Catalog Number: 63-12288

Printed in the United States of America

Current printing (last digit):

10   9   8

TO MY PARENTS

PREFACE

This monograph requires no prior knowledge of combinatorial mathematics. In Chapter 1 we deal with the elementary properties of sets and define *permutation, combination,* and *binomial coefficient*. Of course we treat these concepts from a mature point of view, and from the outset we assume an appreciation for the subtleties of mathematical reasoning. Combinatorial mathematics is best studied within the framework of modern algebra, and for this reason we presuppose a certain familiarity with a few algebraic concepts. Matrices are the really important tool. They occur throughout the monograph and unify the various chapters. At first they are primarily rectangular arrays and little is needed in the way of background. Later they play a fuller role, and we apply the standard rules of matric manipulation. Number theory is used sparingly. An understanding of integral congruences is adequate for most purposes. Groups and fields are mentioned in passing. Only on rare occasions do we call for something beyond the definitions of these systems.

Many of our proofs rely on counting arguments, finite induction, or some other time-tested device. But this does not mean that combinatorial mathematics is easy. The subject is demanding and its exposition is troublesome. Our definitions and proofs are concise and they deserve careful scrutiny. But effort and ingenuity lead to mastery, and our subject holds rich rewards for those who learn its secrets.

We pursue certain topics with thoroughness and reach the frontiers of present-day research. But we pay a price for this and must omit much that is of interest. Each chapter contains its separate bibliography. These are guides for further study and do not aim at completeness. We also discuss in the pages that follow some vital questions that remain unanswered. Combinatorial mathematics is tremendously alive at this moment, and we believe that its greatest truths are still to be revealed.

HERBERT J. RYSER

Syracuse University
February 1963

ACKNOWLEDGMENTS

A number of mathematicians aided me in one way or another. Professor Marshall Hall, Jr. provided considerable help and encouragement. Professors R. P. Boas, Ivan Niven, and Robert Silverman deserve credit for many improvements in the exposition. The following mathematicians must also be mentioned for their varied but valuable contributions: Dr. D. R. Fulkerson, Dr. Karl Goldberg, Dr. A. J. Hoffman, Professor Erwin Kleinfeld, Mr. D. E. Knuth, Dr. Morris Newman, Dr. E. T. Parker, Professor Tibor Rado, and Professor A. W. Tucker. Professor H. M. Gehman and Mr. F. H. Ryser supplied able assistance in matters pertaining to publication. The monograph was written during the period in which I was connected with Ohio State University, California Institute of Technology, U. S. Army Research Office (Durham), and Rand Corporation.

CONTENTS

CHAPTER	PAGE
1. Fundamentals of Combinatorial Mathematics	1
1. What is combinatorial mathematics?	1
2. Sets	3
3. Samples	5
4. Unordered selections	7
5. Binomial coefficients	12
References for Chapter 1	16
2. The Principle of Inclusion and Exclusion	17
1. A fundamental formula	17
2. Applications to number theory	19
3. Derangements	22
4. The permanent	24
References for Chapter 2	28
3. Recurrence Relations	29
1. Some elementary recurrences	29
2. Ménage numbers	31
3. Latin rectangles	35
References for Chapter 3	37
4. A Theorem of Ramsey	38
1. A fundamental theorem	38
2. Applications	43
References for Chapter 4	46
5. Systems of Distinct Representatives	47
1. A fundamental theorem	47
2. Partitions	50
3. Latin rectangles	52
4. Matrices of zeros and ones	53
5. Term rank	55
References for Chapter 5	59

CONTENTS

6. Matrices of Zeros and Ones ... 61

1. The class $\mathfrak{A}(R, S)$ ... 61
2. An application to Latin rectangles ... 65
3. Interchanges ... 67
4. Maximal term rank ... 70
5. Related problems ... 76
   References for Chapter 6 ... 78

7. Orthogonal Latin Squares ... 79

1. Existence theorems ... 79
2. The Euler conjecture ... 84
3. Finite projective planes ... 89
4. Projective planes and Latin squares ... 92
   References for Chapter 7 ... 94

8. Combinatorial Designs ... 96

1. The (b, v, r, k, λ)-configuration ... 96
2. The (v, k, λ)-configuration ... 102
3. A nonexistence theorem ... 108
4. The matric equation $AA^T = B$ ... 116
5. Extremal problems ... 122
   References for Chapter 8 ... 127

9. Perfect Difference Sets ... 131

1. Perfect difference sets ... 131
2. The multiplier theorem ... 134
   References for Chapter 9 ... 141

List of Notation ... 143

Index ... 147

CHAPTER 1

FUNDAMENTALS
OF COMBINATORIAL MATHEMATICS

1. What is combinatorial mathematics? Combinatorial mathematics, also referred to as combinatorial analysis or combinatorics, is a mathematical discipline that began in ancient times. According to legend the Chinese Emperor Yu (c. 2200 B.C.) observed the magic square

$$\begin{bmatrix} 4 & 9 & 2 \\ 3 & 5 & 7 \\ 8 & 1 & 6 \end{bmatrix}$$

on the back of a divine tortoise. Permutations had a feeble beginning in China before 1100 B.C., and Rabbi Ben Ezra (c. 1140 A.D.) seems to have known the formula for the number of combinations of n things taken r at a time. Much of the earliest work is tied to number mysticism. But during the last few centuries various writers have approached the subject from the standpoint of mathematical recreations. Bachet's problem of the weights, Kirkman's schoolgirls problem, and Euler's 36 officers problem are famous illustrations. Such problems are intellectually stimulating, and their solutions are sometimes ingenious and elegant.

Many of the problems studied in the past for their amusement or aesthetic appeal are of great value today in pure and applied science. Not long ago finite projective planes were regarded as a combinatorial curiosity. Today they are basic in the foundations of geometry and in the analysis and design of experiments. Our new technology with its vital concern with the discrete has given the recreational mathematics of the past a new seriousness of purpose.

But more important, the modern era has uncovered for combinatorics a wide range of fascinating new problems. These have arisen in abstract algebra, topology, the foundations of mathematics, graph theory, game theory, linear programming, and in many other areas. Combinatorics has always been diversified. During our day this diversification has increased manyfold. Nor are its many and varied problems successfully attacked in terms of a unified theory. Much of what we have said up to now applies with equal force to the theory of numbers. In fact, combinatorics and number theory are sister disciplines. They share a certain intersection of common knowledge, and each genuinely enriches the other.

Combinatorial mathematics cuts across the many subdivisions of mathematics, and this makes a formal definition difficult. But by and large it is concerned with the study of the arrangement of elements into sets. The elements are usually finite in number, and the arrangement is restricted by certain boundary conditions imposed by the particular problem under investigation. Two general types of problems appear throughout the literature. In the first the existence of the prescribed configuration is in doubt, and the study attempts to settle this issue. These we call existence problems. In the second the existence of the configuration is known, and the study attempts to determine the number of configurations or the classification of these configurations according to types. These we call enumeration problems. This monograph stresses existence problems, but many enumeration problems appear from time to time.

It may be remarked that the second category of problems is nothing more than a refinement or obvious extension of the first. This is the case. But in practice if the existence of a configuration requires intensive study, then almost nothing can be said about the corresponding enumeration problem. On the other hand, if the enumeration problem is tractable, the corresponding existence problem is usually trivial.

We illustrate these remarks with an elementary example. An 8 by 8 checkerboard has 2 squares from opposite corners removed. There are available 31 dominoes, and each domino covers exactly 2 checkerboard squares. The problem is to cover the entire board with the 31 dominoes. In this problem the existence of a solution is in doubt. We show in fact that such a covering is not possible. For 2 black or 2 white squares are deleted. Thus the board has an unequal number of black and white squares. But a domino placed on the board must cover both a black and a white square. Therefore a complete covering is not possible. Suppose the 2 squares from opposite corners are not deleted. Then it is possible to cover the board with 32 dominoes in many ways. Under these circumstances one is led to the enumeration problem of determining the number of distinct coverings.

2. Sets. Let S be an arbitrary set of elements a, b, c, ... We indicate the fact that s is an element of S by writing $s \in S$. If each element of a set A is an element of the set S, then A is a *subset* of S and we designate this by the notation $A \subseteq S$. If $A \subseteq S$ and $S \subseteq A$, then the two sets are identical and we write $A = S$. If $A \subseteq S$ but $A \neq S$, then A is a *proper subset* of S and we write $A \subset S$. The set of all subsets of S is denoted by $P(S)$. For notational convenience the vacuous set or *null set* $\emptyset$ is counted as a member of $P(S)$.

Let S and T be subsets of a set M. The set of elements e such that $e \in S$ and $e \in T$ is called the *intersection* $S \cap T$ of S and T. More generally, if $T_1, T_2, \ldots, T_r$ are subsets of M, then $T_1 \cap T_2 \cap \cdots \cap T_r$ denotes the set of elements e such that $e \in T_i$ for each $i = 1, 2, \ldots, r$. The subsets S and T of M are *disjoint* provided they have no elements in common. The equation $S \cap T = \emptyset$ indicates that S and T are disjoint. The *union* $S \cup T$ of the subsets S and T of M is the set of elements e such that $e \in S$ or $e \in T$. More generally, if $T_1, T_2, \ldots, T_r$ are subsets of M, then $T_1 \cup T_2 \cup \cdots \cup T_r$ denotes the set of all elements e such that $e \in T_i$ for at least one $i = 1, 2, \ldots, r$. The subsets $T_1, T_2, \ldots, T_r$ of M form a *partition* of M provided $M = T_1 \cup T_2 \cup \cdots \cup T_r$ and $T_i \cap T_j = \emptyset$ for $i \neq j$ $(i, j = 1, 2, \ldots, r)$. The partitions of M are *ordered* if equality of the partitions $M = T_1 \cup T_2 \cup \cdots \cup T_r$ and $M = T'_1 \cup T'_2 \cup \cdots \cup T'_r$ means that $T_i = T'_i$ $(i = 1, 2, \ldots, r)$ and *unordered* if equality of the partitions means that each T_i is equal to some T'_j.

A set S containing only a finite number of elements is called a *finite set*. A finite set is a *set of n elements* provided the number of its elements is n. When this terminology is used we take $n > 0$ and exclude the null set $\emptyset$. *Throughout the monograph we call a set of n elements an n-set.* Thus an r-subset of an n-set means a subset of r elements of a set of n elements. Many counting arguments make extensive use of the following elementary principles.

Let S be an m-set and let T be an n-set. If $S \cap T = \emptyset$, then $S \cup T$ is an $(m + n)$-set. This is the *rule of sum*. The *generalized rule of sum* asserts the following. *If T_i is an n_i-set $(i = 1, 2, \ldots, r)$ and if $M = T_1 \cup T_2 \cup \cdots \cup$*

T_r is a partition of M, then M is an $(n_1 + n_2 + \cdots + n_r)$-set.

Let S and T denote two sets and let (s, t) be an ordered pair with $s \in S$ and $t \in T$. Two pairs (s, t) and (s^*, t^*) are equal if $s = s^*$ and $t = t^*$. The set of all of these ordered pairs is called the *product set* of S and T and is denoted by $S \times T$. Let $M(S, T, n)$ denote a set of ordered pairs of the form (s, t), where s is arbitrary in S but each $s \in S$ is paired with exactly n elements $t \in T$. Distinct elements of S need not be paired with elements of the same n-subset of T. The notation implies T contains at least n elements. Moreover, $M(S, T, n) = S \times T$ if and only if T is an n-set. Now let S be an m-set. Then $M(S, T, n)$ is an (mn)-set. This is the *rule of product*. The *generalized rule of product* asserts the following. *If T_1 is an n_1-set and if $M_2 = M(T_1, T_2, n_2)$, $M_3 = M(M_2, T_3, n_3)$, and finally if $M_r = M(M_{r-1}, T_r, n_r)$, then M_r is an $(n_1 n_2 \ldots n_r)$-set.*

3. Samples. Let S be a set and let

(3.1) $\qquad\qquad (a_1, a_2, \ldots, a_r)$

be an ordered r-tuple of not necessarily distinct elements of S. Two such r-tuples $(a_1, a_2, \ldots, a_r)$ and $(a_1^*, a_2^*, \ldots, a_r^*)$ are equal if $a_i = a_i^*$ $(i = 1, 2, \ldots, r)$. We call (3.1) a *sample* of S. The sample is of *size* r, and we refer to (3.1) as an *r-sample* of S.

THEOREM 3.1. *The number of r-samples of an n-set is n^r.*

Proof. Let S be an n-set. This theorem is a special case of the generalized rule of product with $T_1 = T_2 = \cdots = T_r = S$ and $n_1 = n_2 = \cdots = n_r = n$.

Let S be an n-set and let the components a_i of the r-sample (3.1) be distinct. Then the r-sample is called an *r-permutation* of n elements. An r-permutation must have $r \leq n$. An n-permutation is called a *permutation* of n elements.

THEOREM 3.2. *The number of r-permutations of n elements is*

(3.2) $\qquad P(n, r) = n(n - 1) \cdots (n - r + 1).$

Proof. This theorem is a special case of the generalized rule of product with $T_1 = T_2 = \cdots = T_r = S$ and $n_1 = n$, $n_2 = n - 1, \ldots, n_r = n - r + 1$.

By (3.2) $P(n, n)$ stands for the product of the first n positive integers. $P(n, n)$ is called *n-factorial* and is written $n!$. Thus

(3.3) $\qquad P(n, n) = n! = n(n - 1) \cdots 1.$

COROLLARY 3.3. *The number of permutations of n elements is $n!$.*

A (single-valued) *mapping* α of a set S *into* a set T is a correspondence that associates with each $s \in S$ a unique $t = s\alpha \in T$. The element $s\alpha$ is called the *image* of s under the mapping α. Two mappings α and β of S into T are equal if $s\alpha = s\beta$ for all $s \in S$. The mapping α is a mapping of S *onto* T if every $t \in T$ occurs as an image of some $s \in S$. The mapping of S onto T is 1-1 if distinct elements of S have distinct images. Now let $G(S)$ be the set of all 1-1 mappings of S onto itself. Let α and β be in $G(S)$. Then the mapping that sends $s \in S$ into $(s\alpha)\beta \in S$ is a 1-1 mapping called the *product* of the mappings α and β. $G(S)$ is now an algebraic system with a binary composition called product, and one may verify that $G(S)$ satisfies the axioms of a group.

Let S be an n-set of elements labeled $1, 2, \ldots, n$. Then $G(S)$ is called the *symmetric group* of *degree* n. It is denoted by S_n. Let α be the element in S_n that sends i into $i\alpha$ $(i = 1, 2, \ldots, n)$. The 1-1 mapping α is characterized by the permutation

(3.4) $\qquad\qquad (1\alpha, 2\alpha, \ldots, n\alpha).$

Conversely, each permutation of the n elements is in effect a 1-1 mapping of the elements onto themselves. The number of elements in a group is called its *order*. We may now state Corollary 3.3 in the terminology of group theory.

COROLLARY 3.4. *S_n is of order $n!$.*

Examples. (a) The number of 2-permutations of 3 elements is $P(3, 2) = 3 \cdot 2 = 6$. If the elements are labeled 1, 2, 3, the 2-permutations are

$$(1, 2), (1, 3), (2, 1), (2, 3), (3, 1), (3, 2).$$

(b) The number of 5-letter words that may be constructed out of the English alphabet is 26^5. The number of 5-letter words on distinct letters is

$$26 \cdot 25 \cdot 24 \cdot 23 \cdot 22 = 7{,}893{,}600.$$

(c) S_{100} is of order $(9.3326 \ldots)10^{157}$. Eddington's estimate of the number of electrons in the universe is a mere $(136)2^{256}$.

(d) Let A be a matrix of m rows and n columns, and let the entries of A be the integers 0 and 1. There are 2^{mn} of these matrices. If $m = n = 100$, this gives $2^{10,000}$ matrices.

4. Unordered selections. Let S be a set and let

$$(4.1) \qquad \{a_1, a_2, \ldots, a_r\}$$

be an unordered collection of r not necessarily distinct elements of S. The number of occurrences of an element in the collection is called the *multiplicity* of the element. Two such collections $\{a_1, a_2, \ldots, a_r\}$ and $\{a_1^*, a_2^*, \ldots, a_r^*\}$ are equal provided the elements with their respective multiplicities are the same for both collections. We call (4.1) an *unordered selection* of S. The unordered selection is of *size r*, and we refer to (4.1) as an *r-selection* of S. If each element in (4.1) is of multiplicity 1, then the r-selection is an r-subset of S. An r-subset of an n-set is also called an *r-combination* of n elements.

For n a positive integer (3.3) asserts that

(4.2) $$n! = P(n, n).$$

It is convenient to define

(4.3) $$0! = 1$$

so that for every positive integer n

(4.4) $$n! = n(n-1)!.$$

Now let n and r be positive integers and define

(4.5)
$$C(n, r) = \binom{n}{r} = \frac{n(n-1) \cdots (n-r+1)}{r!},$$
$$C(n, 0) = \binom{n}{0} = 1,$$
$$C(0, r) = \binom{0}{r} = 0,$$
$$C(0, 0) = \binom{0}{0} = 1.$$

Then (4.5) defines $C(n, r)$ for all nonnegative integers n and r. Note that if $r > n$, then $C(n, r) = 0$. This means that if n is fixed, $C(n, r)$ takes on only a finite number of distinct values. The numbers $C(n, r)$ defined by (4.5) are the familiar *binomial coefficients*. They are of fundamental importance in enumeration problems.

THEOREM 4.1. *The number of r-subsets of an n-set is*

$$\binom{n}{r}.$$

Proof. By Theorem 3.2 the number of r-permutations of n elements is $P(n, r)$. Each r-permutation may be

Sec. 4   UNORDERED SELECTIONS   9

ordered in $r!$ ways. If order is disregarded, the number of distinguishable arrangements is

$$(4.6) \qquad C(n, r) = \frac{P(n, r)}{r!}.$$

Let S be an n-set and let $P(S)$ denote the set of all subsets of S. Let T denote the set of all n-samples obtained from the 2-set of the integers 0 and 1. There exists a natural 1-1 mapping of $P(S)$ onto T. Thus if $X = \{a_{i_1}, a_{i_2}, \ldots, a_{i_r}\}$ is in $P(S)$, the image of X under the 1-1 mapping is the n-sample with 1's in components $i_1, i_2, \ldots, i_r$ and 0's elsewhere. Now we use Theorem 4.1 to count the elements in $P(S)$ and Theorem 3.1 to count the elements in T. Then if we equate the counts, we obtain

$$(4.7) \qquad \binom{n}{0} + \binom{n}{1} + \cdots + \binom{n}{n} = 2^n.$$

Of course (4.7) is an elementary identity. But the proof of (4.7) illustrates a technique that is effective in many combinatorial investigations.

THEOREM 4.2. *The number of r-selections of an n-set is*

$$(4.8) \qquad \binom{n + r - 1}{n - 1} = \binom{n + r - 1}{r}.$$

Proof. Replace the n-set S by the n-set S' of the positive integers $1, 2, \ldots, n$. Then every r-selection of S' may be written in the form

$$(4.9) \qquad \{a_1, a_2, \ldots, a_r\},$$

where

$$(4.10) \qquad a_1 \leqq a_2 \leqq \cdots \leqq a_r.$$

Now let S^* be the $(n + r - 1)$-set of the positive integers

$1, 2, \ldots, n + r - 1$. Then

(4.11) $\quad \{a_1 + 0, a_2 + 1, \ldots, a_r + r - 1\}$

is an r-subset of S^*. Moreover, the correspondence

(4.12)
$$\{a_1, a_2, \ldots, a_r\} \leftrightarrow \{a_1 + 0, a_2 + 1, \ldots, a_r + r - 1\}$$

is 1-1 from the r-selections of S' onto the r-subsets of S^*. But by Theorem 4.1 the number of r-subsets of S^* is

$$\binom{n + r - 1}{r}.$$

Let

(4.13) $\qquad S = T_1 \cup T_2 \cup \cdots \cup T_k$

be a partition of an n-set S into r_i-subsets T_i ($i = 1, 2, \ldots, k$). Then

(4.14) $\qquad n = r_1 + r_2 + \cdots + r_k,$

and we call the partition (4.13) an $(r_1, r_2, \ldots, r_k)$-*partition* of S.

THEOREM 4.3. *The number of ordered $(r_1, r_2, \ldots, r_k)$-partitions of an n-set is*

(4.15) $$\frac{n!}{r_1! r_2! \ldots r_k!}.$$

Proof. By Theorem 4.1 and the generalized rule of product, the number of ordered $(r_1, r_2, \ldots, r_k)$-partitions of an n-set is

(4.16) $\binom{n}{r_1}\binom{n - r_1}{r_2} \cdots \binom{n - r_1 - \cdots - r_{k-1}}{r_k}$

$$= \frac{n!}{r_1! r_2! \ldots r_k!}.$$

The numbers (4.15) are the *multinomial coefficients*. Theorem 4.3 implies that the number of ordered (1, 1, ..., 1)-partitions of an n-set is $n!$. In this case Theorem 4.3 reduces to Corollary 3.3. The number of ordered $(r, n - r)$-partitions of an n-set is

$$\frac{n!}{r!(n - r)!}.$$

In this case Theorem 4.3 reduces to Theorem 4.1.

The multinomial coefficients of Theorem 4.3 have a second, very useful combinatorial interpretation. Let S be a k-set of elements $a_1, a_2, \ldots, a_k$ and let

(4.17) $$n = r_1 + r_2 + \cdots + r_k,$$

where the r_i are positive integers. Let $(a_{i_1}, a_{i_2}, \ldots, a_{i_n})$ denote an n-sample of S and let this sample contain a_i exactly r_i times $(i = 1, 2, \ldots, k)$. Let T denote the set of all such samples. Then the number of elements in T is

(4.18) $$\frac{n!}{r_1! r_2! \ldots r_k!}.$$

For the r_1 elements a_1 in each sample may be replaced by r_1 distinct elements and distinct from the elements of the sample. Then these r_1 elements may be permuted in $r_1!$ ways, and in this way each sample yields $r_1!$ new samples. Then we apply this replacement procedure to the new collection of samples and replace each of the r_2 elements a_2 by r_2 distinct elements and distinct from the elements of the sample. This procedure terminates with a set of $n!$ permutations, and hence the number of elements in T is given by (4.18).

Examples. (a) A bridge hand consists of a selection of 13 cards from a full deck of 52 cards. The order within a hand is of no concern. Hence the number of different hands is

$$\binom{52}{13} = 635{,}013{,}559{,}600.$$

(b) At bridge the 52 cards of a deck are distributed among 4 players. Each player receives 13 cards. Hence the number of different situations at a bridge table is

$$\frac{52!}{(13!)^4} = (5.3645\ldots)10^{28}.$$

(c) The number of 12-letter words that can be formed from 4 a's, 4 b's, 2 c's, and 2 d's is

$$\frac{12!}{4!4!2!2!} = 207{,}900.$$

(d) The number of 5-letter words that can be formed from the letters a, b, c in which a appears at most twice, b at most once, and c at most three times is

$$\frac{5!}{2!0!3!} + \frac{5!}{2!1!2!} + \frac{5!}{1!1!3!} = 60.$$

(e) A throw with a set of r dice may be regarded as an r-selection of a 6-set. The number of distinct throws is

$$\binom{r+5}{5} = \binom{r+5}{r}.$$

5. Binomial coefficients. Let n and r be positive integers. Then

(5.1) $$\binom{n}{r} = \binom{n-1}{r} + \binom{n-1}{r-1}.$$

Formula (5.1) provides a basic recursion for the binomial coefficients and is an immediate consequence of the definitions (4.5). The investigations of § 4 tell us that the binomial coefficients are integers. We now prove this assertion by induction. If $n = 0$ or if $r = 0$, then the result is valid by (4.5). If n and r are positive, the induction hypothesis implies that the two terms on the right side of

Sec. 5   BINOMIAL COEFFICIENTS

equation (5.1) are integers. Consequently the term on the left side must be an integer. This assertion may be stated in a more striking form. *The product of r successive positive integers is divisible by $r!$.*

We use the term *prime* for a positive integer $\neq 1$ that is divisible only by itself and 1.

THEOREM 5.1. *If p is a prime, then*

(5.2) $$\binom{p}{1}, \binom{p}{2}, \ldots, \binom{p}{p-1}$$

are divisible by p.

Proof. Let r be an integer in the interval $1 \leqq r \leqq p - 1$. Then $r!$ divides

(5.3) $$p(p-1) \cdots (p-r+1).$$

But $r!$ is prime to p and hence $r!$ divides

(5.4) $$(p-1)(p-2) \cdots (p-r+1).$$

Thus

(5.5) $$\binom{p}{r} = p \frac{(p-1)(p-2) \cdots (p-r+1)}{r!}$$

is divisible by p.

Formula (5.1) dictates an effective procedure for the calculation of the binomial coefficients. This is schematically illustrated by a diagram

(5.6)
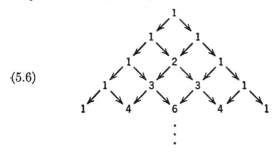

known as the *Pascal triangle*. We regard an arrow in (5.6) as a one-way path. If it is possible to pass from one binomial coefficient P to another binomial coefficient Q in (5.6) along a succession of one-way paths that begin with P and terminate with Q, we say that P and Q are connected by a one-way route. Let I denote the topmost 1 in the triangle. Then I and P are connected by various one-way routes. The binomial coefficient P designates the number of distinct routes. This interesting feature of the Pascal triangle is an inherent property of its construction from (5.1). The symmetry and the monotonicity of the horizontal rows in (5.6) are consequences of the relationships

$$(5.7) \qquad \binom{n}{r} = \binom{n}{n-r} \qquad (0 \leq r \leq n),$$

$$(5.8) \qquad \binom{2n}{0} < \binom{2n}{1} < \cdots < \binom{2n}{n},$$

$$(5.9) \qquad \binom{2n-1}{0} < \binom{2n-1}{1} < \cdots < \binom{2n-1}{n-1} = \binom{2n-1}{n}.$$

Let n be a positive integer. Then

$$(5.10) \quad (x+y)^n = \binom{n}{0}x^n + \binom{n}{1}x^{n-1}y + \cdots + \binom{n}{n}y^n.$$

The algebraic identity (5.10) is the well-known binomial theorem. We give a proof of (5.10) based on the ideas of §4. Let S be an n-set of symbols

$$(5.11) \qquad (x+y)_1, (x+y)_2, \ldots, (x+y)_n.$$

Then for $r > 0$ the coefficient of $x^{n-r}y^r$ in the expansion of $(x+y)^n$ is equal to the number of r-subsets of S. But by Theorem 4.1 this is

(5.12) $$\binom{n}{r}.$$

Hence (5.10) is valid.

Many identities on binomial coefficients are easy consequences of (5.1) and (5.10). Formula (5.1) is ideal for proof by induction of a given identity. The expansion (5.10) is subject to formal manipulation and is a direct source of many relationships among the coefficients. For example, if in (5.10) we set $x = y = 1$, then

(5.13) $$\binom{n}{0} + \binom{n}{1} + \cdots + \binom{n}{n} = 2^n.$$

On the other hand, if we set $x = 1$ and $y = -1$, then

(5.14) $$\binom{n}{0} - \binom{n}{1} + \cdots + (-1)^n \binom{n}{n} = 0.$$

The following identities are typical of those that occur throughout the literature. They may be derived by elementary methods.

(5.15) $$\sum_{k=0}^{n} \binom{n}{k}^2 = \binom{2n}{n},$$

(5.16) $$\sum_{k=1}^{n} k \binom{n}{k} = n \cdot 2^{n-1},$$

(5.17) $$\sum_{k=1}^{n} k^2 \binom{n}{k} = n(n+1) \cdot 2^{n-2},$$

(5.18) $$\sum_{k=1}^{n} \frac{(-1)^{k-1}}{k} \binom{n}{k} = 1 + \frac{1}{2} + \cdots + \frac{1}{n}.$$

References for Chapter 1

Divisibility properties of binomial and multinomial coefficients are discussed in Dickson [1], Chapter 9.

1. L. E. Dickson, *History of the Theory of Numbers*, vol. 1, New York, Chelsea, 1952.

2. W. Feller, *Probability Theory and Its Applications*, vol. 1, New York, Wiley, 1950.

3. E. Netto, *Lehrbuch der Combinatorik*, Leipzig, Teubner, 2nd edition, 1927, reprinted by Chelsea.

4. J. Riordan, *An Introduction to Combinatorial Analysis*, New York, Wiley, 1958.

CHAPTER **2**

THE PRINCIPLE OF INCLUSION AND EXCLUSION

1. A fundamental formula. Let S be an n-set. Let each $a \in S$ be assigned a unique *weight* $w(a)$ with $w(a)$ an element of a field F. No restrictions are placed on the field F or on the weight assignments within F. But frequently the combinatorial problem under consideration suggests some natural weight assignment. In many problems each $a \in S$ is assigned a weight $w(a)$ equal to the positive integer 1. Let P denote an N-set of N properties

(1.1) $$P_1, P_2, \ldots, P_N$$

connected with the elements of S and let

(1.2) $$\{P_{i_1}, P_{i_2}, \ldots, P_{i_r}\}$$

denote an r-subset of P. Let

(1.3) $$W(P_{i_1}, P_{i_2}, \ldots, P_{i_r})$$

equal the sum of the weights of those elements of S that satisfy each of the properties $P_{i_1}, P_{i_2}, \ldots, P_{i_r}$. If no element of S satisfies each of these properties, the expression (1.3) is assigned the value 0. Now let

(1.4) $$W(r) = \Sigma\, W(P_{i_1}, P_{i_2}, \ldots, P_{i_r})$$

equal the sum of the quantities (1.3) over all of the r-subsets of P. We extend (1.4) to the case $r = 0$ and let $W(0)$ equal the sum of the weights of the elements of S. We are now in a position to state the basic inclusion and exclusion formula.

THEOREM 1.1. *Let $E(m)$ denote the sum of the weights of the elements of S that satisfy exactly m of the properties* (1.1). *Then*

$$(1.5) \quad E(m) = W(m) - \binom{m+1}{m} W(m+1) \\ + \binom{m+2}{m} W(m+2) - \cdots + (-1)^{N-m} \binom{N}{m} W(N).$$

Proof. Let $a \in S$ and let a of weight $w(a)$ satisfy exactly t of the properties (1.1). If $t < m$, then a contributes 0 to the right side of (1.5). On the other hand, if $t = m$, then a contributes $w(a)$ to the right side of (1.5). Now if $t > m$, then a contributes

$$(1.6) \quad \left[\binom{t}{m} - \binom{m+1}{m}\binom{t}{m+1} + \binom{m+2}{m}\binom{t}{m+2} - \cdots \right. \\ \left. + (-1)^{t-m} \binom{t}{m}\binom{t}{t} \right] w(a)$$

to the right side of (1.5). But

$$(1.7) \quad \binom{k}{m}\binom{t}{k} = \binom{t}{m}\binom{t-m}{t-k} \quad (m \leq k \leq t),$$

and hence (1.6) reduces to

$$(1.8) \quad \binom{t}{m}\left[\binom{t-m}{t-m} - \binom{t-m}{t-(m+1)} \right. \\ \left. + \binom{t-m}{t-(m+2)} - \cdots + (-1)^{t-m}\binom{t-m}{t-t}\right] w(a).$$

But by (5.14) of Chapter 1 the bracketed expression in (1.8) is equal to 0. Hence if $t > m$, then a contributes 0 to the right side of (1.5). This implies that the right side of (1.5) is the sum of the weights of the elements of S that satisfy exactly m of the properties (1.1).

THEOREM 1.2. *Let $E(0)$ denote the sum of the weights of the elements of S that satisfy none of the properties* (1.1). *Then*

(1.9)
$$E(0) = W(0) - W(1) + W(2) - \cdots + (-1)^N W(N).$$

Proof. This is the case $m = 0$ of Theorem 1.1.

Let each $a \in S$ be assigned a weight $w(a)$ equal to the positive integer 1. Then a sum of weights is the number of entries in the sum. Theorem 1.2 specialized in this way has $W(0) = n$, and $E(0)$ is the number of elements of S that satisfy none of the properties (1.1). Equation (1.9) specialized in this way is called the *sieve formula*. It is attributed to da Silva and Sylvester. Actually the sieve formula is very old, and it may have been known in one form or another to the Bernoullis. We devote the remaining sections in this chapter to various applications of Theorem 1.1.

2. Applications to number theory. In this section we apply the sieve formula to selected topics in elementary number theory. If x is a real number ≥ 0, let

(2.1) $$[x]$$

denote the greatest integer $\leq x$. Let

(2.2) $$(a, b)$$

denote the positive g. c. d. of two integers a and b not both 0. Thus $(a, b) = 1$ means that a and b are relatively prime. We write

(2.3) $\qquad a \mid b$

if a divides b and

(2.4) $\qquad a \nmid b$

if a does not divide b.

THEOREM 2.1. *Let n be a positive integer and let $a_1, a_2, \ldots, a_N$ be positive integers such that $(a_i, a_j) = 1$ for $i \neq j$. Then the number of integers k such that*

(2.5) $\quad 0 < k \leq n, \qquad a_i \nmid k \qquad (i = 1, 2, \ldots, N),$

is

(2.6) $\quad n - \sum_{1 \leq i \leq N} \left[\dfrac{n}{a_i}\right] + \sum_{1 \leq i < j \leq N} \left[\dfrac{n}{a_i a_j}\right] - \cdots$
$$+ (-1)^N \left[\dfrac{n}{a_1 a_2 \ldots a_N}\right].$$

Proof. Let S be the n-set of positive integers $1, 2, \ldots, n$ and let P_i be the property that an element of S is divisible by a_i ($i = 1, 2, \ldots, N$). Now the a_i are relatively prime in pairs. Hence the expression

(2.7) $\qquad W(P_{i_1}, P_{i_2}, \ldots, P_{i_r})$

in the sieve formula is the number of integers k such that

(2.8) $\qquad 0 < k \leq n, \qquad a_{i_1} a_{i_2} \ldots a_{i_r} \mid k.$

But this number is

(2.9) $\qquad \left[\dfrac{n}{a_{i_1} a_{i_2} \ldots a_{i_r}}\right].$

The *Euler φ-function* $\varphi(n)$ of the positive integer n is the number of integers k such that

(2.10) $\qquad 0 < k \leq n, \qquad (k, n) = 1.$

THEOREM 2.2. *Let n be a positive integer. Then*

$$(2.11) \quad \varphi(n) = n \prod_p \left(1 - \frac{1}{p}\right).$$

The product in (2.11) extends over all prime divisors p of n.

Proof. In Theorem 2.1 replace a_i by p_i and suppose that $p_1, p_2, \ldots, p_N$ are the prime divisors of n. Then (2.6) implies

$$(2.12) \quad \varphi(n) = n - \sum_{1 \leq i \leq N} \frac{n}{p_i} + \sum_{1 \leq i < j \leq N} \frac{n}{p_i p_j} - \cdots + (-1)^N \frac{n}{p_1 p_2 \ldots p_N}.$$

But this is equivalent to (2.11).

The *Möbius function* $\mu(n)$ of the positive integer n is defined by

$$\mu(1) = 1,$$

$(2.13) \quad \mu(n) = 0 \quad$ if n is divisible by the square of a prime,

$$\mu(p_1 p_2 \ldots p_k) = (-1)^k \quad \text{if the primes } p_1, p_2, \ldots, p_k \text{ are distinct.}$$

The Möbius function allows us to write (2.12) in a more elegant form:

$$(2.14) \quad \varphi(n) = n \sum_d \frac{\mu(d)}{d}.$$

The summation in (2.14) extends over all positive divisors d of n.

Let n be a positive integer. If the primes $\leq \sqrt{n}$ are known, then the primes $\leq n$ may be found as follows. We write the sequence of integers

$$(2.15) \quad 2, 3, \ldots, n.$$

We then strike out from (2.15) all numbers divisible by 2, then all numbers divisible by 3, then all numbers divisible by 5, and so on, and finally all numbers divisible by q, where q is the largest prime $\leq \sqrt{n}$. The numbers that remain are all of the primes that are $> \sqrt{n}$ and $\leq n$, for a number that remains cannot have a prime factor $\leq \sqrt{n}$, nor can it be the product of two numbers $> \sqrt{n}$. This effective procedure for the construction of primes is called the *sieve of Eratosthenes*.

Now let x be a positive real and let $\pi(x)$ denote the number of primes $\leq x$. After the application of the sieve of Eratosthenes to the sequence (2.15) there remain exactly

$$(2.16) \qquad \pi(n) - \pi(\sqrt{n})$$

integers. But the number of integers that remain may be calculated in a second way. In Theorem 2.1 replace a_i by q_i and suppose that $q_1, q_2, \ldots, q_N$ are all of the primes $\leq \sqrt{n}$. Then by Theorem 2.1 the number in question is

$$(2.17) \quad -1 + n - \sum_{1 \leq i \leq N} \left[\frac{n}{q_i}\right] + \sum_{1 \leq i < j \leq N} \left[\frac{n}{q_i q_j}\right] - \cdots$$
$$+ (-1)^N \left[\frac{n}{q_1 q_2 \cdots q_N}\right].$$

Hence it follows that

$$(2.18) \quad \pi(n) - \pi(\sqrt{n}) = -1 + \sum_d \mu(d) \left[\frac{n}{d}\right].$$

The summation in (2.18) extends over all positive divisors d of the product $q_1 q_2 \cdots q_N$, where $q_1, q_2, \ldots, q_N$ are the primes $\leq \sqrt{n}$.

3. Derangements. Let

$$(3.1) \qquad (a_1, a_2, \ldots, a_n)$$

be a permutation of n elements labeled $1, 2, \ldots, n$. The permutation (3.1) is a *derangement* if $a_i \neq i$ ($i = 1, 2,$

$\ldots, n$). Thus a derangement has no element in its natural position. A problem of Montmort, commonly known by its French name, "le problème des recontres," asks for the number of these derangements. Let D_n denote this number. We may evaluate D_n without difficulty by the sieve formula. For let S be the set of the $n!$ permutations (3.1) and let P_i be the property that the permutation (3.1) has $a_i = i$ ($i = 1, 2, \ldots, n$). Then

$$(3.2) \qquad W(P_{i_1}, P_{i_2}, \ldots, P_{i_r}) = (n - r)!$$

and

$$(3.3) \qquad W(r) = \binom{n}{r}(n-r)! = \frac{n!}{r!}.$$

Thus we obtain the following formula for D_n:

$$(3.4) \quad D_n = n!\left(1 - \frac{1}{1!} + \frac{1}{2!} - \cdots + (-1)^n \frac{1}{n!}\right).$$

Formula (3.4) brings to mind the alternating series for e^{-1}

$$(3.5) \qquad e^{-1} = 1 - \frac{1}{1!} + \frac{1}{2!} - \cdots.$$

In fact we may now write (3.5) in the form

$$(3.6) \qquad e^{-1} = \frac{D_n}{n!} + (-1)^{n+1}\frac{1}{(n+1)!} \pm \cdots,$$

and this means that $D_n/n!$ and e^{-1} differ by less than $1/(n+1)!$. Hence $n!e^{-1}$ is a very good approximation for D_n.

Formula (3.4) has a number of rather amusing applications. Suppose, for example, that n gentlemen attend a party and place their n hats in a checkroom. Thereupon the hats are mixed and returned at random to the guests. The probability that no gentleman receives his own hat is equal to $D_n/n!$. This application has many variants.

But the surprising feature is that the probability is for all practical purposes e^{-1}, and it matters hardly at all if 10 or 10,000 gentlemen are present. A problem with more serious overtones asks for the number of ways in which 8 rooks may be placed on a conventional chessboard so that no rook can attack another and the white diagonal is free of rooks. This may be done in $D_8 = 14{,}833$ distinct ways.

4. The permanent. In what follows we assume an acquaintanceship with the elementary portions of the theory of matrices, and we introduce at this point a notation and terminology that will be in use throughout the volume. Let S be a set. A *rectangular array* based on the set S is a configuration of m rows and n columns of the form

$$(4.1) \qquad A = \begin{bmatrix} a_{11} & a_{12} & \cdots & a_{1n} \\ a_{21} & a_{22} & \cdots & a_{2n} \\ \cdot & \cdot & & \cdot \\ \cdot & \cdot & & \cdot \\ \cdot & \cdot & & \cdot \\ a_{m1} & a_{m2} & \cdots & a_{mn} \end{bmatrix}.$$

The entry a_{ij} in row i and column j of A must be an element of S, but the set S need not be restricted in any way. We say that a_{ij} occupies the (i, j) *position* of A. Whenever we wish to stress the fact that A contains m rows and n columns, we refer to A as an m by n array or, equivalently, we say that A is of *size m by n*. In case $m = n$, then A is a *square* array and A is of *order n*. If $m - r$ rows and $n - s$ columns of A are deleted, the resulting rectangular array of size r by s is called a *subarray* of A. Two m by n arrays are equal if corresponding entries in the (i, j) positions are equal $(i = 1, 2, \ldots, m; j = 1, 2, \ldots, n)$. In a certain sense the array (4.1) is nothing more than a sample of size mn of the given set S. But from another standpoint a 1 by n array may be regarded as a sample of size n, and thus

(4.1) is a natural generalization of this concept. We now replace the rather cumbersome notation of (4.1) by

(4.2) $\quad\quad A = [a_{ij}] \quad (i = 1, 2, \ldots, m; j = 1, 2, \ldots, n).$

Let $e = \min(m, n)$. Then the *main diagonal* of A consists of the entries a_{ii} in positions (i, i) for $i = 1, 2, \ldots, e$. The *transpose* A^T of A is the n by m array obtained from A by a reflection of A about its main diagonal. Thus A^T contains a_{ji} in its (i, j) position $(i = 1, 2, \ldots, n; j = 1, 2, \ldots, m)$. The array A is *symmetric* provided $A = A^T$.

Suppose now that the set S is a field F. Then the rectangular array is a matrix. Addition and scalar multiplication for m by n matrices may be defined in the usual way, and the set of all m by n matrices with elements in F constitutes a vector space of dimension mn over F. Moreover, an m by n matrix may be multiplied by an n by t matrix under the familiar row by column rule for the multiplication of matrices. The resulting matrix is of size m by t. We note that if A is of size m by n, we may always form the products AA^T of order m and A^TA of order n. These products are in fact symmetric matrices.

Now let $A = [a_{ij}]$ be a matrix of size m by n with $m \leq n$. Then we define the *permanent* of A by

(4.3) $\quad\quad \text{per}(A) = \Sigma\, a_{1i_1} a_{2i_2} \ldots a_{mi_m}.$

The summation in (4.3) extends over all m-permutations $(i_1, i_2, \ldots, i_m)$ of the integers $1, 2, \ldots, n$. This scalar function of the matrix A appears repeatedly in the literature of combinatorics in connection with certain enumeration and extremal problems. We now consider some of the formal properties of per (A). In the first place per (A) remains invariant under arbitrary permutations of the rows and the columns of A. Also, the multiplication of a row of A by a scalar a in F replaces per (A) by $a \cdot \text{per}(A)$. Next we consider the important case in which A is a square matrix of order n. Then per (A) is also invariant under transposition, and we may write

(4.4) $$\text{per } (A) = \text{per } (A^T).$$

In this case per (A) is the same as the determinant det (A) apart from a factor ± 1 preceding each product on the right side of equation (4.3). This suggests the possibility of a computational procedure for per (A) analogous to the well-developed theory for det (A). As a matter of fact, certain determinantal laws have analogues for permanents. For example, the Laplace expansion for determinants has a simple counterpart for permanents. But the basic multiplicative law valid for determinants

(4.5) $$\det (AB) = \det (A) \det (B)$$

is flagrantly false for permanents. Also, the addition of a multiple of one row of A to another does not leave per (A) invariant. These facts greatly inhibit computational techniques for per (A), and consequently many square matrices have easily evaluated determinants and undetermined permanents. We now describe a procedure for the evaluation of per (A).

THEOREM 4.1. *Let A be a matrix of size m by n with $m \leq n$. Let A_r denote a matrix obtained from A by replacing r columns of A by zeros. Let $S(A_r)$ denote the product of the row sums of A_r and let $\Sigma S(A_r)$ denote the sums of the $S(A_r)$ over all of the choices for A_r. Then*

(4.6)
$$\text{per } (A) = \Sigma S(A_{n-m}) - \binom{n-m+1}{1} \Sigma S(A_{n-m+1})$$
$$+ \binom{n-m+2}{2} \Sigma S(A_{n-m+2}) - \cdots$$
$$+ (-1)^{m-1} \binom{n-1}{m-1} \Sigma S(A_{n-1}).$$

Proof. Let S denote the set of all samples of size m

$$(4.7) \qquad (j_1, j_2, \ldots, j_m)$$

of the positive integers $1, 2, \ldots, n$. Let the weight of the sample (4.7) equal

$$(4.8) \qquad a_{1j_1} a_{2j_2} \ldots a_{mj_m}.$$

Let P_i be the property that the sample (4.7) does not contain the integer i ($i = 1, 2, \ldots, n$). Now suppose that A_r is obtained from A by replacing the columns numbered $i_1, i_2, \ldots, i_r$ by 0's. Then

$$(4.9) \qquad W(P_{i_1}, P_{i_2}, \ldots, P_{i_r}) = S(A_r)$$

and hence

$$(4.10) \qquad W(r) = \Sigma S(A_r).$$

The function per (A) is equal to the sum of the weights of the elements of S that satisfy exactly $n - m$ of the properties P_i ($i = 1, 2, \ldots, n$). Thus (4.6) is a consequence of Theorem 1.1.

COROLLARY 4.2. *Let A be a square matrix of order n. Then*

$$(4.11) \quad \text{per } (A) = S(A) - \Sigma S(A_1) + \Sigma S(A_2) - \cdots \\ + (-1)^{n-1} \Sigma S(A_{n-1}).$$

Proof. This is Theorem 4.1 with $m = n$.

Let A be a matrix and let the entries of A be the integers 0 and 1. We call such a matrix a (0, 1)-*matrix*. The 2^{mn} (0, 1)-matrices of size m by n are of fundamental importance in combinatorics, and they will play a leading role in our development of the subject. For the moment we confine ourselves to a few elementary remarks concerning the permanents of some very special (0, 1)-matrices. Let I denote the identity matrix of order n and let J denote the matrix of order n with every entry equal to 1. Then it is trivial to verify that

$$\text{(4.12)} \qquad \text{per }(J) = n!$$

and

$$\text{(4.13)} \qquad \text{per }(J - I) = D_n.$$

We remark in conclusion that (4.11) and (4.12) imply the identity

$$\text{(4.14)} \qquad n! = \sum_{r=0}^{n-1} (-1)^r \binom{n}{r} (n - r)^n.$$

Moreover (4.11) and (4.13) yield a second formula for the number of derangements, namely

(4.15)
$$D_n = \sum_{r=0}^{n-1} (-1)^r \binom{n}{r} (n - r)^r (n - r - 1)^{n-r}.$$

References for Chapter 2

1. W. Feller, *Probability Theory and Its Applications*, vol. 1, New York, Wiley, 1950.
2. G. H. Hardy and E. M. Wright, *An Introduction to the Theory of Numbers*, Oxford University Press, 3rd edition, 1954.
3. T. Nagell, *Introduction to Number Theory*, New York, Wiley, 1951.
4. J. Riordan, *An Introduction to Combinatorial Analysis*, New York, Wiley, 1958.

CHAPTER **3**

RECURRENCE RELATIONS

1. Some elementary recurrences. The relationship

$$(1.1) \qquad \binom{n}{r} = \binom{n-1}{r} + \binom{n-1}{r-1}$$

is a simple instance of a recurrence. From (1.1) and the appropriate initial values one may evaluate the binomial coefficients for all nonnegative integers n and r. The evaluation procedure is schematically illustrated by the Pascal triangle. Relationships of many different types and varieties are called recurrences, and there is no need for us to insist on a formal definition of the term. But by and large a recurrence stands for a special type of relationship involving a quantity with integer parameters. This relationship is such that it may be used to evaluate the quantity term by term from given initial values and from previously computed values. Recurrences arise naturally in many enumeration problems, and the theory of recurrences has an extensive literature. The subject is developed with great thoroughness in the recent treatise by Riordan cited at the end of this chapter. We do not pursue this topic apart from a discussion of a few of the simpler recurrences of special interest to us.

We begin with a problem from elementary geometry. Suppose that we ask for the number of subdivisions into which the plane is divided by n straight lines in general position. Let P_n denote this number. We define

(1.2) $$P_0 = 1.$$

Then it is easy to verify that for every positive integer n

(1.3) $$P_n = P_{n-1} + n.$$

The initial condition (1.2) and the recurrence (1.3) determine P_n for all nonnegative integers n. In fact (1.2) and (1.3) imply

(1.4) $$P_n = \frac{n(n+1)}{2} + 1.$$

Consider next the set T of all samples of size n obtained from the 2-set of the integers 0 and 1. We ask for the number of samples that do not contain two successive 0's. Let $f(n)$ denote this number. We define

(1.5) $$f(0) = 1,$$

and it is trivial that

(1.6) $$f(1) = 2.$$

Now let $n \geq 2$. Then there are $f(n-1)$ such samples with the first component equal to 1 and $f(n-2)$ such samples with the first component equal to 0. Hence

(1.7) $$f(n) = f(n-1) + f(n-2).$$

The initial conditions (1.5) and (1.6) and the recurrence (1.7) determine $f(n)$ for all nonnegative integers n. The numbers $f(n)$ are called *Fibonacci numbers*. They have many remarkable arithmetical and combinatorial properties.

Euler investigated derangements from the standpoint of recurrences. We define

(1.8) $$D_0 = 1,$$

and it is trivial that

(1.9) $$D_1 = 0.$$

Consider now a derangement

(1.10) $$(a_1, a_2, \ldots, a_n)$$

of n elements labeled 1, 2, ..., n with $n \geq 2$. The first position in (1.10) is open to all elements except 1 and hence to $n - 1$ elements. Suppose that the first entry of (1.10) is fixed with $a_1 = k$ ($k \neq 1$). Then the derangements (1.10) are of two types according to whether 1 is or is not in the kth position. If 1 is in the kth position, then the number of permutations is that of $n - 2$ elements with all elements displaced, and this is D_{n-2}. On the other hand, if 1 is not in the kth position, then the permutations permitted are those that involve the elements 1, 2, ..., $k - 1$, $k + 1$, ..., n in positions 2 through n with 1 not in the kth position and every other element out of its own position. But this is the same as the permutations of $n - 1$ elements labeled 2 through n with every element displaced. Hence the number of these permutations is D_{n-1}. The preceding remarks imply

(1.11) $$D_n = (n - 1)(D_{n-1} + D_{n-2}).$$

The recurrence (1.11) may be used to give a straightforward proof by induction of the formula

(1.12) $$D_n = n!\left(1 - \frac{1}{1!} + \frac{1}{2!} - \cdots + (-1)^n \frac{1}{n!}\right).$$

2. Ménage numbers. Let U_n denote the number of permutations of n elements labeled 1, 2, ..., n such that the element i is in neither of the positions i and $i + 1$ for $i = 1, 2, \ldots, n - 1$, and the element n is in neither of the positions n and 1. In other words, U_n is the number of permutations discordant with the two permutations

(2.1) $$(1, 2, 3, \ldots, n), \qquad (n, 1, 2, \ldots, n - 1).$$

Let C denote the $(0, 1)$-matrix of order n with 1's in the positions $(1, 2), (2, 3), (3, 4), \ldots, (n, 1)$ and 0's elsewhere. Let J denote the matrix of order n with every entry equal to 1 and let I denote the identity matrix of order n. Then it follows without difficulty that

$$(2.2) \qquad U_n = \text{per } (J - I - C).$$

But the formula for the permanent in the preceding chapter does not lead to an immediate evaluation of U_n.

The numbers U_n are called *ménage numbers*. The reason for this is the following "problème des ménages" formulated by Lucas. This asks for the number of ways of seating n married couples at a circular table with men and women in alternate positions and such that no wife sits next to her husband. The wives may be seated first, and this may be done in $2n!$ ways. Then each husband is excluded from the two seats beside his wife, but the number of ways of seating the husbands is independent of the seating arrangement of the wives. Thus if M_n denotes the number of seating arrangements for the "problème des ménages," it is clear that

$$(2.3) \qquad M_n = 2n! U_n.$$

Consequently we may concentrate our attention on the ménage numbers U_n.

THEOREM 2.1. *The ménage numbers U_n are given by*

$$(2.4) \quad U_n = n! - \frac{2n}{2n-1}\binom{2n-1}{1}(n-1)!$$
$$+ \frac{2n}{2n-2}\binom{2n-2}{2}(n-2)! - \cdots$$
$$+ (-1)^n \frac{2n}{n}\binom{n}{n} 0! \qquad (n > 1).$$

This remarkable formula for U_n was announced in a communication by Touchard. Our proof of (2.4) is an elegant recurrence argument of Kaplansky.

LEMMA 2.2. *Let $f(n, k)$ denote the number of ways of selecting k objects, no two consecutive, from n objects arranged in a row. Then*

$$(2.5) \qquad f(n, k) = \binom{n - k + 1}{k}.$$

Proof. We have the initial conditions

$$(2.6) \qquad f(n, 1) = \binom{n}{1} = n$$

and for $n > 1$,

$$(2.7) \qquad f(n, n) = \binom{1}{n} = 0.$$

Now let $1 < k < n$. We may split the selections into those that include the first object and those that do not. The selections that include the first object cannot include the second object and are enumerated by

$$(2.8) \qquad f(n - 2, k - 1).$$

The selections that do not include the first object are enumerated by

$$(2.9) \qquad f(n - 1, k).$$

Hence we have the recurrence

$$(2.10) \quad f(n, k) = f(n - 1, k) + f(n - 2, k - 1).$$

We may now prove (2.5) by induction. Our induction hypothesis asserts that

$$(2.11)$$
$$f(n - 1, k) = \binom{n - k}{k}, \qquad f(n - 2, k - 1) = \binom{n - k}{k - 1}.$$

But then (2.10) and (2.11) imply

$$(2.12) \qquad f(n, k) = \binom{n-k}{k} + \binom{n-k}{k-1},$$

and this is equivalent to (2.5).

LEMMA 2.3. *Let $g(n, k)$ denote the number of ways of selecting k objects, no two consecutive, from n objects arranged in a circle. Then*

$$(2.13) \qquad g(n, k) = \frac{n}{n-k}\binom{n-k}{k} \qquad (n > k).$$

Proof. As before we may split the selections into those that include the first object and those that do not. The selections that include the first object cannot include the second object or the last object and are enumerated by

$$(2.14) \qquad f(n-3, k-1).$$

The selections that do not include the first object are enumerated by

$$(2.15) \qquad f(n-1, k).$$

Hence

$$(2.16) \quad g(n, k) = f(n-1, k) + f(n-3, k-1),$$

and (2.13) is now an easy consequence of Lemma 2.2.

We return once again to the permutations on the elements labeled $1, 2, \ldots, n$. Let P_i be the property that a permutation has i in position i ($i = 1, 2, \ldots, n$) and let P_i' be the property that a permutation has i in position $i + 1$ ($i = 1, 2, \ldots, n-1$) with P_n' the property that the permutation has n in position 1. Now let the $2n$ properties be listed in a row

$$(2.17) \qquad P_1, P_1', P_2, P_2', \ldots, P_n, P_n'.$$

We select k of these properties and ask for the number of permutations that satisfy each of the k properties. The answer is $(n-k)!$ if the k properties are compatible and 0 otherwise. Let v_k denote the number of ways of selecting k compatible properties from the $2n$ properties (2.17). Then by the sieve formula

$$(2.18) \quad U_n = v_0 n! - v_1(n-1)! + v_2(n-2)! - \cdots \\ + (-1)^n v_n 0!.$$

It remains to evaluate v_k. But we note that if the $2n$ properties (2.17) are arranged in a circle, then only the consecutive ones are not compatible. Hence by Lemma 2.3,

$$(2.19) \qquad v_k = \frac{2n}{2n-k}\binom{2n-k}{k},$$

and this gives us our main conclusion.

3. Latin rectangles. Let S be a set of n elements. A *Latin rectangle* based on the n-set S is an r by s rectangular array

$$(3.1) \qquad A = [a_{ij}] \quad (i = 1, 2, \ldots, r; j = 1, 2, \ldots, s)$$

with the requirement that each row of (3.1) is an s-permutation of elements of S and each column of (3.1) is an r-permutation of elements of S. This implies that we must have $r \leq n$ and $s \leq n$. Let the elements of S be labeled $1, 2, \ldots, n$ and suppose that $s = n$. Then the Latin rectangle contains a permutation of the elements $1, 2, \ldots, n$ in each row, and the permutations are chosen so that no column contains repeated elements. A Latin rectangle of this type is *normalized* provided that the first row is written in standard order $1, 2, \ldots, n$. If $L(r, n)$ denotes the number of r by n Latin rectangles and if $K(r, n)$ denotes the number of normalized r by n Latin rectangles, then trivially

$$(3.2) \qquad L(r, n) = n! K(r, n).$$

Normalized 2 by n Latin rectangles are derangements and consequently

(3.3) $$K(2, n) = D_n.$$

It is also clear that the ménage numbers U_n are the number of 3 by n Latin rectangles with the initial rows

(3.4) $$\begin{bmatrix} 1 & 2 & 3 & \cdots & n \\ n & 1 & 2 & \cdots & n-1 \end{bmatrix}.$$

An interesting formula of Riordan for $K(3, n)$ asserts that

(3.5) $$K(3, n) = \sum_{k=0}^{m} \binom{n}{k} D_{n-k} D_k U_{n-2k},$$

where in (3.5), $m = [n/2]$ and $U_0 = 1$. But the enumeration of Latin rectangles of more than three lines has scarcely been touched. In this connection mention should be made of an important asymptotic formula of Erdös and Kaplansky. This asserts that if $r < (\log n)^{3/2}$, then

(3.6) $$L(r, n) \sim n!^r e^{-\binom{r}{2}}.$$

They conjectured that (3.6) remains valid for $r < n^{1/3}$, and this has been established by Yamamoto.

If $r = s = n$, then the Latin rectangle is called a *Latin square* of *order* n. Such configurations may be regarded as multiplication tables of very general algebraic systems. We point out that the multiplication table of a finite group defines a Latin square. But a Latin square constructed in this way possesses very special properties. If now

(3.7) $$L(n, n) = n!(n-1)!l_n,$$

then l_n denotes the number of Latin squares of order n with the first row and column in standard order. Our previous remarks concerning $K(r, n)$ suggest that the evaluation of

l_n may not be easy. This is in fact the case, and the following table displays the known values of l_n.

n	1	2	3	4	5	6	7
l_n	1	1	1	4	56	9408	16,942,080

References for Chapter 3

Riordan [5] gives a thorough account of recurrences with many additional references to the literature. Fibonnaci numbers are discussed in Dickson [1], Chapter 17.

Classical papers on ménage numbers include Touchard [7] and Kaplansky [3]. The proof of the formula for $K(3, n)$ appears in Riordan [4]. Asymptotic formulas for $L(r, n)$ are discussed by Erdös and Kaplansky [2] and Yamamoto [8]. The value of l_7 is due to Sade [6].

1. L. E. Dickson, *History of the Theory of Numbers*, vol. 1, New York, Chelsea, 1952.

2. P. Erdös and I. Kaplansky, The asymptotic number of Latin rectangles, *Amer. Jour. Math.*, 68 (1946), 230–236.

3. I. Kaplansky, Solution of the "Problème des ménages," *Bull. Amer. Math. Soc.*, 49 (1943), 784–785.

4. J. Riordan, Three-line Latin rectangles-II, *Amer. Math. Monthly*, 53 (1946), 18–20.

5. ———, *An Introduction to Combinatorial Analysis*, New York, Wiley, 1958.

6. A. Sade, *Énumération des Carrés Latins. Application au 7ᵉ Ordre. Conjecture pour les Ordres Supérieurs*, Marseille, 1948.

7. J. Touchard, Sur un problème de permutations, *C. R. Acad. Sci. Paris*, 198 (1934), 631–633.

8. K. Yamamoto, On the asymptotic number of Latin rectangles, *Japanese Jour. Math.*, 21 (1951), 113–119.

CHAPTER **4**

A THEOREM OF RAMSEY

1. A fundamental theorem. We devote this section to the description and proof of an important combinatorial theorem that originated in certain investigations in the foundations of mathematics. The theorem is called Ramsey's theorem after the English logician F. P. Ramsey. The pigeon-hole principle in mathematics asserts that if a set of sufficiently many elements is partitioned into not too many subsets, then at least one of the subsets must contain many of the elements. Ramsey's theorem may be regarded as a profound generalization of this simple principle. The proof of Ramsey's theorem makes extensive use of recurrence techniques. The theorem has a wide variety of applications, and some of these are discussed in the following section.

Let S be an n-set and let $P_r(S)$ be the set of all r-subsets of S. Let

$$(1.1) \qquad P_r(S) = A_1 \cup A_2 \cup \cdots \cup A_t$$

be an arbitrary ordered partition of $P_r(S)$ into t components $A_1, A_2, \ldots, A_t$. Now let $q_1, q_2, \ldots, q_t$ be integers such that

$$(1.2) \qquad 1 \leqq r \leqq q_1, q_2, \ldots, q_t.$$

Sec. 1   A FUNDAMENTAL THEOREM   39

If there exists a q_i-subset of S with all of its r-subsets in A_i, we call the subset a (q_i, A_i)-*subset* of S. Ramsey's theorem asserts the following.

THEOREM 1.1. *Let the given integers $q_1, q_2, \ldots, q_t$, and r satisfy (1.2). Then there exists a minimal positive integer $N(q_1, q_2, \ldots, q_t, r)$ with the property that the following proposition is valid for all integers $n \geq N(q_1, q_2, \ldots, q_t, r)$. Let S be an n-set and let (1.1) be an arbitrary ordered partition of $P_r(S)$ into t components $A_1, A_2, \ldots, A_t$. Then S contains a (q_i, A_i)-subset for some $i = 1, 2, \ldots, t$.*

Proof. We first specialize Ramsey's theorem in various ways in order to obtain a better understanding of the statement of the theorem. We begin by noting that for $r = 1$ the theorem reduces to the pigeon-hole principle. For in this case $P_r(S)$ is S, and a (q_i, A_i)-subset of S is a q_i-subset of A_i. Hence it follows that

(1.3)
$$N(q_1, q_2, \ldots, q_t, 1) = q_1 + q_2 + \cdots + q_t - t + 1.$$

Next we set $q_1 = q_2 = \cdots = q_t = q \geq r \geq 1$. The theorem specialized in this way asserts the following. Let S be an n-set with n sufficiently large. Let the r-subsets of S be partitioned into t components in an arbitrary manner. Then there exists a q-subset of S with all of its r-subsets in one of the t components. Actually it is easy to derive Ramsey's theorem in its full generality from this assertion. This may be accomplished by selecting q equal to the maximum of the integers $q_1, q_2, \ldots, q_t$. Ramsey's theorem is trivial for $t = 1$, and in this case we have $N(q_1, r) = q_1$. Suppose for the moment that we have established the theorem for $t = 2$. Then we show that the theorem must be valid for $t = 3$. For we may write

(1.4) $$P_r(S) = A_1 \cup (A_2 \cup A_3),$$

and we may let

(1.5) $$q_2' = N(q_2, q_3, r).$$

Now if $n \geq N(q_1, q_2', r)$, then the n-set S must contain a (q_1, A_1)-subset or a $(q_2', A_2 \cup A_3)$-subset. But if the latter alternative holds, then the q_2'-subset of S must contain a (q_2, A_2)-subset or a (q_3, A_3)-subset. Consequently the theorem is valid for $t = 3$, and it is clear that by induction the theorem is valid for all t. Consequently we are required to prove the theorem only for $t = 2$, and we now confine our attention to this case.

By (1.3) we know that

(1.6) $$N(q_1, q_2, 1) = q_1 + q_2 - 1.$$

Moreover, we assert that

(1.7) $$N(q_1, r, r) = q_1,$$

(1.8) $$N(r, q_2, r) = q_2.$$

To prove this let $q_2 = r$ and let $n \geq q_1$. If A_2 is not empty, then the n-set S contains an (r, A_2)-subset. On the other hand, if A_2 is empty, then $A_1 = P_r(S)$ and S contains a $(q_1, P_r(S))$-subset. This proves (1.7). Equation (1.8) follows by a similar argument.

We now use induction to complete the proof for $t = 2$. By (1.6), (1.7), and (1.8) we may assume that the given integers q_1, q_2, and r satisfy $1 < r < q_1, q_2$. We take as our induction hypothesis the existence of the integers $N(q_1 - 1, q_2, r)$ and $N(q_1, q_2 - 1, r)$. Moreover, we include in the induction hypothesis the existence of the integer $N(q_1', q_2', r - 1)$ for all q_1', q_2' such that $1 \leq r - 1 \leq q_1', q_2'$. From these assumptions and from (1.6), (1.7), and (1.8) we establish the existence of the integer $N(q_1, q_2, r)$. This is a valid induction, for then the following integers exist.

$$N(2, 2, 2), \quad N(2, 3, 2), \quad N(2, 4, 2), \ldots$$
$$(1.9) \quad N(3, 2, 2), \quad N(3, 3, 2), \quad N(3, 4, 2), \ldots$$
$$N(4, 2, 2), \quad N(4, 3, 2), \quad N(4, 4, 2), \ldots$$

$$\begin{matrix} \cdot & \cdot & \cdot \\ \cdot & \cdot & \cdot \\ \cdot & \cdot & \cdot \end{matrix}$$

This in turn implies the existence of the integers $N(q_1, q_2, 3)$, and so on.

By the induction hypothesis we are assured of the existence of $p_1 = N(q_1 - 1, q_2, r)$, $p_2 = N(q_1, q_2 - 1, r)$, and $N(p_1, p_2, r - 1)$. We establish the existence of $N(q_1, q_2, r)$. In fact, we prove the recurrence inequality

$$(1.10) \quad N(q_1, q_2, r) \leqq N(p_1, p_2, r - 1) + 1.$$

Let

$$(1.11) \quad n \geqq N(p_1, p_2, r - 1) + 1.$$

Let a be a fixed element of the n-set S and let T be the $(n - 1)$-set of the elements of S with a excluded. We use the partition $P_r(S) = A_1 \cup A_2$ of all r-subsets of S to define a partition

$$(1.12) \quad P_{r-1}(T) = B_1 \cup B_2$$

of the set $P_{r-1}(T)$ of all $(r - 1)$-subsets of T. This is done as follows. Let R be an arbitrary $(r - 1)$-subset of T. If $R \cup a$ is in A_1, we place R in B_1, and if $R \cup a$ is in A_2, we place R in B_2. This gives us the partition (1.12) of $P_{r-1}(T)$.

The set T contains at least $N(p_1, p_2, r - 1)$ elements. Hence T contains a (p_1, B_1)-subset or a (p_2, B_2)-subset.

We deal first with the case in which T contains a (p_1, B_1)-subset. Then T contains a p_1-subset U, all of whose $(r-1)$-subsets are in B_1. Now $p_1 = N(q_1 - 1, q_2, r)$, and U as a subset of S contains a $(q_1 - 1)$-subset, all of whose r-subsets are in A_1 or a q_2-subset, all of whose r-subsets are in A_2. If the latter alternative holds, then the q_2-subset satisfies our requirements and we are finished. Suppose then that U contains a $(q_1 - 1)$-subset V, all of whose r-subsets are in A_1. But in this case $W = V \cup a$ is a q_1-subset of S. If an r-subset of W does not contain a, then it is an r-subset of V, and hence the r-subset is in A_1. On the other hand, if an r-subset of W contains a, then it consists of a and an $(r-1)$-subset of V. The set V as a subset of U has all of its $(r-1)$-subsets in B_1. But then by the definition of the partition (1.12) such an r-subset of W is also in A_1. Thus W is a q_1-subset of S with all of its r-subsets in A_1. The alternative case in which T contains a (p_2, B_2)-subset may be treated by an entirely analogous argument. This completes the proof of Ramsey's theorem.

The integers $N(q_1, q_2, r)$ have a deep combinatorial significance. But unfortunately no recurrence is known for these integers, and the recurrence inequality (1.10) is not very sharp in most instances. This causes serious difficulties in the evaluation of $N(q_1, q_2, r)$. Of course we always have the trivial values (1.6), (1.7), and (1.8). But apart from these all known $N(q_1, q_2, r)$ are contained in the following symmetric array for $N(q_1, q_2, 2)$.

(1.13)

	3	4	5
3	6	9	14
4	9	18	
5	14		

It is not surprising that even less is known for $t > 2$. In this case the main piece of information at present is the assertion

(1.14) $$N(3, 3, 3, 2) = 17.$$

2. Applications. Consider n points in general position in three-dimensional space. Two distinct points determine a line segment, and let each of these line segments be colored either red or blue. The 2-subsets of points may then be partitioned into the set A_1 of red segments and into the set A_2 of blue segments. Now if q_1 and q_2 are integers such that $2 \leq q_1, q_2$ and if $n \geq N(q_1, q_2, 2)$, then Ramsey's theorem assures us that there must exist q_1 points with all segments red or q_2 points with all segments blue. Moreover, $N(q_1, q_2, 2)$ is the minimal integer with this property.

Our next application of Ramsey's theorem concerns convex polygons.

THEOREM 2.1. *Let m be an integer greater than or equal to three. Then there exists a minimal positive integer N_m such that the following proposition is valid for all integers $n \geq N_m$. If n points in the plane have no three points collinear, then m of the points are the vertices of a convex m-gon.*

LEMMA 2.2. *If five points in the plane have no three points collinear, then four of the points are the vertices of a convex quadrilateral.*

Proof. The five points determine ten line segments, and the perimeter of this configuration is a convex polygon. If this convex polygon is a pentagon or a quadrilateral, the lemma is trivial. Suppose that the convex polygon is a triangle. Then two of the five points are in the interior of the triangle. The two interior points determine a straight line, and two of the three points of the triangle lie on one side of this line. Then these two points of the triangle and the two interior points form a convex quadrilateral.

LEMMA 2.3. *If m points in the plane have no three points collinear and if all quadrilaterals formed from the m points are convex, then the m points are the vertices of a convex m-gon.*

Proof. The m points determine $m(m-1)/2$ line segments, and the perimeter of this configuration is a convex q-gon. Let the consecutive vertices of the q-gon be labeled $V_1, V_2, \ldots, V_q$. If one of our points is within the q-gon, it must lie in one of the triangles $V_1V_2V_3$, $V_1V_3V_4$, $\ldots$, $V_1V_{q-1}V_q$. But this contradicts the assertion that all quadrilaterals formed from the m points are convex. Hence $q = m$ and the m-gon is convex.

Theorem 2.1 is now an easy consequence of Ramsey's theorem. To prove this let $m \geq 4$ and let $n \geq N(5, m, 4)$. We partition the 4-subsets of the n points into the concave and the convex quadrilaterals. Then by Ramsey's theorem there exists a 5-gon with all quadrilaterals concave or an m-gon with all quadrilaterals convex. By Lemma 2.2 the first alternative cannot occur, and by Lemma 2.3 the m-gon is convex.

Our argument has shown that

$$(2.1) \qquad N_m \leq N(5, m, 4).$$

We know that $N_3 = 3 = 2 + 1$, $N_4 = 5 = 2^2 + 1$, and it has been shown that $N_5 = 9 = 2^3 + 1$. This leads one to conjecture that

$$(2.2) \qquad N_m = 2^{m-2} + 1,$$

but the assertion (2.2) is an unsettled question.

Our concluding application concerns $(0, 1)$-matrices. A submatrix of order m of a matrix A of order n is called *principal* provided the submatrix is obtained from A by deleting $n - m$ of its rows and the same $n - m$ columns.

Theorem 2.4. *Let m be an arbitrary positive integer. Then every $(0, 1)$-matrix A of a sufficiently large order n contains a principal submatrix of order m of one of the following four types:*

$$(2.3) \quad \begin{bmatrix} * & & & 0 \\ & \cdot & & \\ & & \cdot & \\ 0 & & & * \end{bmatrix}, \begin{bmatrix} * & & & 0 \\ & \cdot & & \\ & & \cdot & \\ 1 & & & * \end{bmatrix},$$

$$\begin{bmatrix} * & & & 1 \\ & \cdot & & \\ & & \cdot & \\ 0 & & & * \end{bmatrix}, \begin{bmatrix} * & & & 1 \\ & \cdot & & \\ & & \cdot & \\ 1 & & & * \end{bmatrix}.$$

The asterisks on the main diagonal denote 0's and 1's, but the entries above the main diagonal and the entries below the main diagonal are all 0's or all 1's as illustrated in (2.3).

Proof. Let the n-set S of Ramsey's theorem be the set of the n row vectors of $A = [a_{ij}]$. Denote row i of A by α_i. We let $i < j$, and we associate with the row vectors α_i and α_j of A the vector (a_{ji}, a_{ij}). Now this vector is $(0, 0)$, $(1, 0)$, $(0, 1)$, or $(1, 1)$. Hence we have a partition of the 2-subsets of S

$$(2.4) \qquad P_2(S) = A_1 \cup A_2 \cup A_3 \cup A_4.$$

Now suppose that

$$(2.5) \qquad n \geq N(m, m, m, m, 2).$$

Then by Ramsey's theorem there exists an m-subset of S with all of its 2-subsets in one of the four components of $P_2(S)$. But this implies the existence of a principal submatrix of one of the four types illustrated in (2.3).

References for Chapter 4

Ramsey's theorem in its original form appears in [8]. Our proof and the results on convex polygons are by Erdös and Szekeres [4]. The computed values for $N(q_1, q_2, 2)$ and $N(3, 3, 3, 2)$ appear in Greenwood and Gleason [6].

1. P. Erdös and R. Rado, A combinatorial theorem, *Jour. London Math. Soc.*, **25** (1950), 249–255.

2. ———, Combinatorial theorems on classifications of subsets of a given set, *Proc. London Math. Soc.*, 3rd series, **2** (1952), 417–439.

3. ———, A partition calculus in set theory, *Bull. Amer. Math. Soc.*, **62** (1956), 427–489.

4. P. Erdös and G. Szekeres, A combinatorial problem in geometry, *Compositio Mathematica*, **2** (1935), 463–470.

5. A. W. Goodman, On sets of acquaintances and strangers at any party, *Amer. Math. Monthly*, **66** (1959), 778–783.

6. R. E. Greenwood and A. M. Gleason, Combinatorial relations and chromatic graphs, *Canad. Jour. Math.*, **7** (1955), 1–7.

7. R. Rado, Direct decomposition of partitions, *Jour. London Math. Soc.*, **29** (1954), 71–83.

8. F. P. Ramsey, On a problem of formal logic, *Proc. London Math. Soc.*, 2nd series, **30** (1930), 264–286.

9. T. Skolem, Ein kombinatorischer Satz mit Anwendung auf ein logisches Entscheidungsproblem, *Fundamenta Mathematicae*, **20** (1933), 254–261.

CHAPTER **5**

SYSTEMS OF DISTINCT REPRESENTATIVES

1. A fundamental theorem. Let S be an arbitrary set and let $P(S)$ denote the set of all subsets of S. Let

$$(1.1) \qquad D = (a_1, a_2, \ldots, a_m)$$

be an m-sample of S and let

$$(1.2) \qquad M(S) = (S_1, S_2, \ldots, S_m)$$

be an m-sample of $P(S)$. Now suppose that the m elements of D are distinct and suppose that

$$(1.3) \qquad a_i \in S_i \qquad (i = 1, 2, \ldots, m).$$

Then the element a_i *represents* the set S_i, and we say that the subsets $S_1, S_2, \ldots, S_m$ have a *system of distinct representatives* (abbreviated SDR). We call D an SDR for $M(S)$. The definition of SDR requires $a_i \neq a_j$ whenever $i \neq j$, but S_i and S_j need not be distinct as subsets of S.

We give a simple illustration of the SDR concept. Let S be the 5-set of the integers 1, 2, 3, 4, 5. Let $S_1 = \{2, 5\}$, $S_2 = \{2, 5\}$, $S_3 = \{1, 2, 3, 4\}$, $S_4 = \{1, 2, 5\}$. Then $D = (2, 5, 3, 1)$ is an SDR for (S_1, S_2, S_3, S_4). In this example if we replace S_4 by $S_4' = \{2, 5\}$, then the subsets no longer have an SDR. For $S_1 \cup S_2 \cup S_4'$ is a 2-set, and three elements are required to represent S_1, S_2, S_4'.

The following theorem of P. Hall gives a necessary and sufficient condition for the existence of an SDR.

THEOREM 1.1. *The subsets $S_1, S_2, \ldots, S_m$ have an SDR if and only if the set $S_{i_1} \cup S_{i_2} \cup \cdots \cup S_{i_k}$ contains at least k elements for $k = 1, 2, \ldots, m$ and for all k-combinations $\{i_1, i_2, \ldots, i_k\}$ of the integers $1, 2, \ldots, m$.*

The necessity of this fundamental theorem is obvious. We now prove a refinement of the sufficiency that gives a lower bound on the number of SDR's. Then in the subsequent sections we discuss further ramifications and applications of our results.

THEOREM 1.2. *Let the subsets $S_1, S_2, \ldots, S_m$ satisfy the necessary condition for the existence of an SDR and let each of these subsets contain at least t elements. If $t \leq m$, then $M(S)$ has at least $t!$ SDR's, and if $t > m$, then $M(S)$ has at least $t!/(t-m)!$ SDR's.*

Proof. The proof is by induction on m. The theorem is trivial for $m = 1$. We take as our induction hypothesis the statement of the theorem for all m'-samples of $P(S)$, where $m' < m$. We prove the theorem for the m-sample $M(S) = (S_1, S_2, \ldots, S_m)$. The proof splits into two cases. In the first case we assume that the set $S_{i_1} \cup S_{i_2} \cup \cdots \cup S_{i_k}$ contains at least $k + 1$ elements. This holds for $k = 1, 2, \ldots, m - 1$ and for all k-combinations $\{i_1, i_2, \ldots, i_k\}$ of the integers $1, 2, \ldots, m$. We dispose of this case as follows. Let a_1 be a fixed element of S_1. Delete a_1 whenever it appears in the sets $S_2, S_3, \ldots, S_m$ and call the resulting sets $S_2', S_3', \ldots, S_m'$, respectively. The $(m-1)$-sample

(1.4) $$M'(S) = (S_2', S_3', \ldots, S_m')$$

satisfies the necessary condition for the existence of an

SDR. This is because the set $S_{i_1} \cup S_{i_2} \cup \cdots \cup S_{i_k}$ contains at least $k + 1$ elements. Now if $t \leq m$, then $t - 1 \leq m - 1$, and by the induction hypothesis $M'(S)$ has at least $(t - 1)!$ SDR's. Also, if $t > m$, then $t - 1 > m - 1$, and by the induction hypothesis $M'(S)$ has at least $(t - 1)!/(t - m)!$ SDR's. But a_1 and an SDR for $M'(S)$ give us an SDR for $M(S)$ in which a_1 represents S_1. This holds for each of the t choices for a_1. Hence we obtain the desired number of SDR's for $M(S)$.

In the second case there exists a k-subset of S of the form $S_{i_1} \cup S_{i_2} \cup \cdots \cup S_{i_k}$. Here k is an integer in the interval $1 \leq k \leq m - 1$, and $\{i_1, i_2, \ldots, i_k\}$ is a certain k-combination of the integers $1, 2, \ldots, m$. We renumber the subsets $S_1, S_2, \ldots, S_m$ so that $S_{i_1} \cup S_{i_2} \cup \cdots \cup S_{i_k}$ is $S_1 \cup S_2 \cup \cdots \cup S_k$. The existence of this k-subset implies $t \leq k$. Hence by the induction hypothesis the k-sample $(S_1, S_2, \ldots, S_k)$ has at least $t!$ SDR's. Let $D^* = (a_1, a_2, \ldots, a_k)$ denote one such SDR. Delete elements of D^* whenever they appear in the sets $S_{k+1}, S_{k+2}, \ldots, S_m$ and call the resulting sets $S^*_{k+1}, S^*_{k+2}, \ldots, S^*_m$, respectively. The $(m - k)$-sample

$$(1.5) \qquad M^*(S) = (S^*_{k+1}, S^*_{k+2}, \ldots, S^*_m)$$

satisfies the necessary condition for the existence of an SDR. For if, say, $S^*_{k+1} \cup S^*_{k+2} \cup \cdots \cup S^*_{k+k^*}$ contains fewer than k^* elements, then

(1.6)

$$S_1 \cup S_2 \cup \cdots \cup S_k \cup S_{k+1} \cup S_{k+2} \cup \cdots \cup S_{k+k^*}$$

contains fewer than $k + k^*$ elements, and this is contrary to the hypothesis of the theorem. Hence by the induction hypothesis $M^*(S)$ has at least one SDR and consequently $M(S)$ has at least $t!$ SDR's. This proves Theorem 1.2 and also Theorem 1.1.

2. Partitions. Let

$$(2.1) \qquad T = A_1 \cup A_2 \cup \cdots \cup A_m$$

and

$$(2.2) \qquad T = B_1 \cup B_2 \cup \cdots \cup B_m$$

denote two partitions of a set T with no components equal to the null set $\emptyset$. Let E be an m-subset of T such that each $A_i \cap E \neq \emptyset$ and each $B_j \cap E \neq \emptyset$. Then each of these intersections must be a 1-set and E is called a *system of common representatives* (abbreviated SCR) for the partitions (2.1) and (2.2). An SCR exists for these partitions if and only if there exists a suitable renumbering of the components of (2.1) such that

$$(2.3) \qquad A_i \cap B_i \neq \emptyset \qquad (i = 1, 2, \ldots, m).$$

We use SDR theory to obtain the following necessary and sufficient condition for the existence of an SCR.

THEOREM 2.1. *The partitions* (2.1) *and* (2.2) *have an SCR if and only if the set* $A_{i_1} \cup A_{i_2} \cup \cdots \cup A_{i_k}$ *contains at most k of the components* $B_1, B_2, \ldots, B_m$ *for* $k = 1, 2, \ldots, m$ *and for all k-combinations* $\{i_1, i_2, \ldots, i_k\}$ *of the integers* $1, 2, \ldots, m$.

Proof. The necessity of the theorem is once again obvious. We prove the sufficiency as follows. Let S be the m-set of the elements $A_1, A_2, \ldots, A_m$ and let S_i be the set of all A_j such that $A_j \cap B_i \neq \emptyset$. Then $M(S) = (S_1, S_2, \ldots, S_m)$ is an m-sample of subsets of S. We assert that $M(S)$ satisfies the necessary condition for the existence of an SDR. For if, say, $S_1 \cup S_2 \cup \cdots \cup S_{k+1}$ contains only k elements $A_{i_1}, A_{i_2}, \ldots, A_{i_k}$, then $A_{i_1} \cup A_{i_2} \cup \cdots \cup A_{i_k}$ contains the $k + 1$ components $B_1, B_2, \ldots, B_{k+1}$, and this is contrary to the hypothesis of the theorem. Hence by Theorem 1.1 there exists an SDR for $M(S)$. Now we may renumber the components of (2.1) so that this SDR is $D = (A_1, A_2, \ldots, A_m)$. But then (2.3) is valid and this proves Theorem 2.1.

THEOREM 2.2. *Let $T = A_1 \cup A_2 \cup \cdots \cup A_m$ and $T = B_1 \cup B_2 \cup \cdots \cup B_m$ denote two partitions of T, where each A_i and each B_j is an r-subset of T. Then the partitions have an SCR.*

Proof. This is a special case of Theorem 2.1.

Theorem 2.2 implies the following for r by m arrays based on the integers $1, 2, \ldots, rm$. Let

$$(2.4) \quad A = \begin{bmatrix} 1 & 2 & \cdots & m \\ m+1 & m+2 & \cdots & 2m \\ \cdot & \cdot & & \\ \cdot & \cdot & & \\ \cdot & \cdot & & \\ (r-1)m+1 & (r-1)m+2 & \cdots & rm \end{bmatrix}.$$

Now let B be an r by m array and let the rm entries of B be the integers $1, 2, \ldots, rm$, but in arbitrary positions within B. Then there exists a permutation of the columns of B such that corresponding columns of A and B each contain at least one element in common.

Our next illustration requires an understanding of the elementary properties of cosets in the theory of groups. But the theorem is then a direct consequence of Theorem 2.2.

THEOREM 2.3. *Let G be a finite group and let H be a subgroup of G. Let $G = Hx_1 \cup Hx_2 \cup \cdots \cup Hx_m$ be a right coset decomposition for H and let $G = y_1H \cup y_2H \cup \cdots \cup y_mH$ be a left coset decomposition for H. Then there exist elements $z_1, z_2, \ldots, z_m$ in G such that*

(2.5)

$$G = Hz_1 \cup Hz_2 \cup \cdots \cup Hz_m = z_1H \cup z_2H \cup \cdots \cup z_mH.$$

3. Latin rectangles. In this section we apply SDR theory to Latin rectangles. Let there be given an r by s Latin rectangle based on n elements labeled $1, 2, \ldots, n$. We say that the Latin rectangle may be *extended* to a Latin square of order n provided that we may adjoin $n - r$ rows and $n - s$ columns to the Latin rectangle so that the resulting configuration is a Latin square of order n. The new configuration contains the old in the upper left corner.

THEOREM 3.1. *Let there be given an r by n Latin rectangle based on n elements labeled $1, 2, \ldots, n$. Then the Latin rectangle may be extended to a Latin square of order n.*

Proof. Let S be the n-set of elements $1, 2, \ldots, n$ and let S_i be the set of all elements of S that do not appear in column i of the Latin rectangle. Then each S_i is an $(n - r)$-subset of S, and $M(S) = (S_1, S_2, \ldots, S_n)$ is an n-sample of subsets of S. We prove that $M(S)$ satisfies the necessary condition for the existence of an SDR. Let i be an element of S. The r occurrences of i in the Latin rectangle are confined to distinct columns. Hence the element i is in exactly $n - r$ of the sets $S_1, S_2, \ldots, S_n$. Now if, say, $S_1 \cup S_2 \cup \cdots \cup S_k$ contains only $k - 1$ elements, then these $k - 1$ elements appear in the sets $S_1, S_2, \ldots, S_k$ no more than $(n - r)(k - 1)$ times. But this contradicts the fact that each of these sets is an $(n - r)$-subset of S. Hence $M(S)$ has an SDR. Denote this SDR by $D = (i_1, i_2, \ldots, i_n)$. Then D may be adjoined to the r by n Latin rectangle to yield an $r + 1$ by n Latin rectangle. The entire process may be repeated until the Latin rectangle is extended to a Latin square of order n.

THEOREM 3.2. *There are at least*

(3.1) $$n!(n - 1)! \ldots (n - r + 1)!$$

r by n Latin rectangles and hence at least

(3.2) $$n!(n-1)!\ldots 1!$$

n by n Latin squares.

Proof. There are $n!$ Latin rectangles of size 1 by n. By Theorem 3.1 and Theorem 1.2 each of these may be extended to at least $(n-1)!$ Latin rectangles of size 2 by n. Hence there are at least $n!(n-1)!$ Latin rectangles of size 2 by n. The argument may be repeated and this proves the theorem.

Let l_n denote the number of Latin squares of order n with the first row and column in standard order. Then Theorem 3.2 asserts that

(3.3) $$l_n \geq (n-2)!(n-3)!\ldots 1!.$$

The following table displays the values of l_n and $b_n = (n-2)!(n-3)!\ldots 1!$ for $n = 3, 4, 5, 6, 7$.

n	3	4	5	6	7
l_n	1	4	56	9408	16,942,080
b_n	1	2	12	288	34,560

4. Matrices of zeros and ones. Let A be a matrix of size m by n and let the entries of A be the integers 0 and 1. These $(0, 1)$-matrices play a leading role in the development of many combinatorial topics. One of the chief reasons for this is the following. Let S be an n-set of elements $a_1, a_2, \ldots, a_n$ and let $M(S) = (S_1, S_2, \ldots, S_m)$ be an m-sample of subsets of S. Now let $a_{ij} = 1$ if a_j is a member of S_i and let $a_{ij} = 0$ if a_j is not a member of S_i. Then

(4.1) $$A = [a_{ij}] \quad (i = 1, 2, \ldots, m; j = 1, 2, \ldots, n)$$

is a $(0, 1)$-matrix of size m by n. This matrix is called the *incidence matrix* for the subsets $S_1, S_2, \ldots, S_m$ of the n-set S. The 1's in row i of A specify the elements that belong to set S_i, and the 1's in column j of A specify the sets that contain element a_j. Thus A contains a complete description of the subsets $S_1, S_2, \ldots, S_m$ of S. Also, if a (0, 1)-matrix A of size m by n is given and if S is an arbitrary n-set, there exist subsets $S_1, S_2, \ldots, S_m$ of S such that A is the incidence matrix for these subsets.

From the preceding discussion it is clear that the (0, 1)-matrix A characterizes the subsets $S_1, S_2, \ldots, S_m$ of S. We may characterize these subsets equally well by a matrix with entries $+1$ and -1 or for that matter by a matrix with two distinct entries x and y. These characterizations are entirely satisfactory, but they frequently have no special advantage over the characterization by the incidence matrix A. In fact the choice of 0 and 1 for the entries of A is especially convenient because of the simple behavior of these integers under addition and multiplication. The following theorem illustrates this point.

THEOREM 4.1. *Let $S_1, S_2, \ldots, S_m$ be subsets of an n-set and let $m \leq n$. Let A be the incidence matrix for these subsets. Then the number of SDR's for $M(S) = (S_1, S_2, \ldots, S_m)$ is per (A).*

Proof. The proof is a direct consequence of the terminology. Note that the definition of per (A) and the existence of an SDR both require $m \leq n$.

A *permutation matrix* P is a (0, 1)-matrix of size m by n such that $PP^T = I$, where P^T denotes the transpose of P and I denotes the identity matrix of order m. This definition implies $m \leq n$. A permutation matrix of order m has a single entry 1 in each row and column and all other entries 0. Suppose now that the elements and the subsets of S are renumbered. Then the incidence matrix A is replaced by an incidence matrix A' of the form

(4.2) $$A' = PAQ.$$

Here P is a permutation matrix of order m determined by the renumbering of the subsets, and Q is a permutation matrix of order n determined by the renumbering of the elements. Many investigations involving the (0, 1)-matrix A deal with functions like per (A) that remain invariant under arbitrary permutations of the rows and the columns of A. The reason for this is now apparent. Such functions are of interest in combinatorics because they do not depend on the particular labeling of the elements and the subsets of S.

5. Term rank. A *line* of a matrix designates either a row or a column of the matrix. The *trace* of a matrix is the sum of the entries on the main diagonal of the matrix. Now let A be a (0, 1)-matrix of size m by n. The *term rank* of A is the maximal number of 1's in A with no two 1's on a line. Thus the term rank of A is equal to the maximal trace of A under arbitrary permutations of the rows and the columns of A. This is equivalent to the maximal order of a square submatrix of A with a nonzero permanent. The term rank provides a convenient generalization of the SDR concept for the subsets $S_1, S_2, \ldots, S_m$ of an n-set S. For if A is the incidence matrix for these subsets, then the subsets have an SDR if and only if the term rank of A equals m.

THEOREM 5.1. *Let A be a (0, 1)-matrix of size m by n. The minimal number of lines in A that contain all of the 1's in A is equal to the term rank of A.*

Proof. Let ρ' equal the minimal number of lines in A that contain all of the 1's in A and let ρ equal the term rank of A. We must prove that $\rho = \rho'$. No line can contain two of the 1's that account for the ρ 1's of the term rank. Hence $\rho' \geqq \rho$. We use SDR theory to show that $\rho \geqq \rho'$. Let the minimal covering of 1's by ρ' lines consist of e rows and f columns, where $e + f = \rho'$. Both ρ and ρ' are invariant under permutations of the rows and the columns of A. Hence we take these e rows and f columns as initial rows and columns of the matrix, and we note that $e \leq n - f$. We write the matrix in the form

(5.1) $$\begin{bmatrix} A_1 & A_2 \\ A_3 & A_4 \end{bmatrix},$$

where A_1 is of size e by f. Now A_2 is of term rank e. For we may regard A_2 as an incidence matrix for subsets $S_1, S_2, \ldots, S_e$ of the $(n-f)$-set of the integers $f+1$, $f+2, \ldots, n$. These subsets must satisfy the necessary condition for the existence of an SDR. For if this condition is violated, then we may replace certain of the e rows by fewer columns and retain the covering of 1's in A. But this covering is accomplished with fewer than $e + f$ lines, and this contradicts the minimality of ρ'. We may regard the transpose A_3^T of A_3 as an incidence matrix for subsets and show in the same way that A_3 is of term rank f. Hence $\rho \geq e + f = \rho'$ and this proves Theorem 5.1.

Theorem 5.1 has the following immediate generalization. Let A be a matrix of size m by n with elements in a field F. The minimal number of lines in A that contain all of the nonzero entries in A is equal to the maximal number of nonzero entries in A with no two nonzero entries on a line.

THEOREM 5.2. *Let A be a matrix of size m by n. Let the entries of A be nonnegative reals and let $m \leq n$. Let each row sum of A equal m' and let each column sum of A equal n'. Then*

(5.2) $$A = c_1 P_1 + c_2 P_2 + \cdots + c_t P_t,$$

where in (5.2) each P_i is a permutation matrix and each c_j is a nonnegative real.

Proof. If A is not a square matrix, we replace A by

$$(5.3) \qquad A' = \begin{bmatrix} A \\ \frac{m'}{n} J \end{bmatrix},$$

where J is the matrix of 1's of size $n - m$ by n. The matrix A' is of order n, and the entries of A' are nonnegative reals. Each row and column sum of A' is equal to m'. If A' is not the zero matrix, A' has n positive entries with no two on a line. For if A' did not have n such entries, then by the remarks following Theorem 5.1 we could cover the positive entries in A' with e rows and f columns, where $e + f < n$. But then $m'n \leq m'(e + f) < m'n$, and this is a contradiction. Now let P_1' be the permutation matrix of order n with 1's in the same positions occupied by the n positive entries of A'. Let c_1 be the smallest of these n entries. Then $A' - c_1 P_1'$ is a matrix whose entries are nonnegative reals. Also, $A' - c_1 P_1'$ has each row and column sum equal to the nonnegative real $m' - c_1$. But at least one more zero entry appears in $A' - c_1 P_1'$ than in A'. Hence we may now work on $A' - c_1 P_1'$, and we may repeat the argument until $A' = c_1 P_1' + c_2 P_2' + \cdots + c_t P_t'$. But this gives us a decomposition of the form (5.2) for the matrix A.

Theorem 5.2 has a number of interesting applications.

Theorem 5.3. *Let A be a $(0, 1)$-matrix of order n such that each row and column sum of A is equal to the positive integer k. Then*

$$(5.4) \qquad A = P_1 + P_2 + \cdots + P_k,$$

where the P_i are permutation matrices.

Proof. This follows from the proof of Theorem 5.2. We have each $c_j = 1$ and the process terminates in $t = k$ steps.

Theorem 5.3 gives an affirmative answer to the following problem. A dance is attended by n boys and n girls. Each boy has been previously introduced to exactly k girls and each girl has been previously introduced to exactly k boys. No one desires to make further introductions. Can the boys and the girls be paired so that no further introductions are necessary? Let $A = [a_{ij}]$ be the (0, 1)-matrix defined by $a_{ij} = 1$ if boy j has been previously introduced to girl i and 0 otherwise. Then A satisfies the requirements of Theorem 5.3, and the permutation matrix P_1 of (5.4) gives the desired pairing of boys and girls.

A matrix A of order n is called *doubly stochastic* provided its entries are nonnegative reals and its row and column sums are each equal to 1. These matrices have been extensively studied in their own right because of their importance in the theory of transition probabilities. Theorem 5.2 implies the following.

THEOREM 5.4. *Let A be a doubly stochastic matrix of order n. Then*

(5.5) $$A = c_1 P_1 + c_2 P_2 + \cdots + c_t P_t,$$

where the P_i are permutation matrices and the c_j are positive reals such that

(5.6) $$c_1 + c_2 + \cdots + c_t = 1.$$

Let A be doubly stochastic. The entries of A are nonnegative reals so per (A) cannot exceed the product of the row sums of A. But since each row sum of A is 1, we have

(5.7) $$\text{per } (A) \leq 1.$$

Equality holds in (5.7) if and only if the doubly stochastic A is a permutation matrix. By Theorem 5.4 it is clear that if A is doubly stochastic, then per $(A) > 0$. But if A

is doubly stochastic of order n, then the determination of the minimal value of per (A) is a difficult unsolved problem. A conjecture of van der Waerden asserts

$$(5.8) \qquad \text{per } (A) \geqq \frac{n!}{n^n}.$$

Equality holds in (5.8) if $A = n^{-1}J$, where J is the matrix of 1's of order n. In fact this may be the only case of equality. But up to the present time the validity of the van der Waerden conjecture has been verified for only the first few values of n.

References for Chapter 5

The fundamental Theorem 1.1 is by P. Hall [8]. Theorem 1.2 is by M. Hall [6]. Our proof of Theorem 1.2 follows Halmos and Vaughan [9] and Mann and Ryser [15]. The applications to Latin rectangles are by M. Hall [5; 6]. Many of the applications to matrices are by König [14]. Extensive studies on the van der Waerden conjecture appear in Marcus and Newman [17; 18].

1. C. Berge, *Théorie des Graphes et Ses Applications*, Paris, Dunod, 1958.
2. C. J. Everett and G. Whaples, Representations of sequences of sets, *Amer. Jour. Math.*, **71** (1949), 287–293.
3. L. R. Ford, Jr., and D. R. Fulkerson, Network flows and systems of representatives, *Canad. Jour. Math.*, **10** (1958), 78–85.
4. ———, *Flows in Networks*, Princeton University Press, 1962.
5. M. Hall, Jr., An existence theorem for Latin squares, *Bull. Amer. Math. Soc.*, **51** (1945), 387–388.
6. ———, Distinct representatives of subsets, *Bull. Amer. Math. Soc.*, **54** (1948), 922–926.
7. ———, An algorithm for distinct representatives, *Amer. Math. Monthly*, **63** (1956), 716–717.
8. P. Hall, On representatives of subsets, *Jour. London Math. Soc.*, **10** (1935), 26–30.
9. P. R. Halmos and H. E. Vaughan, The marriage problem, *Amer. Jour. Math.*, **72** (1950), 214–215.

10. P. J. Higgins, Disjoint transversals of subsets, *Canad. Jour. Math.*, **11** (1959), 280–285.

11. A. J. Hoffman, Some recent applications of the theory of linear inequalities to extremal combinatorial analysis, *Proc. of Symposia in Applied Math.*, **10** (1960), 113–128.

12. A. J. Hoffman and H. W. Kuhn, Systems of distinct representatives and linear programming, *Amer. Math. Monthly*, **63** (1956), 455–460.

13. ———, On systems of distinct representatives, *Annals of Math. Studies*, no. 38 (1956), 199–206.

14. D. König, *Theorie der Endlichen und Unendlichen Graphen*, New York, Chelsea, 1950.

15. H. B. Mann and H. J. Ryser, Systems of distinct representatives, *Amer. Math. Monthly*, **60** (1953), 397–401.

16. M. Marcus and H. Minc, On the relation between the determinant and the permanent, *Illinois Jour. Math.*, **5** (1961), 376–381.

17. M. Marcus and M. Newman, On the minimum of the permanent of a doubly stochastic matrix, *Duke Math. Jour.*, **26** (1959), 61–72.

18. ———, Inequalities for the permanent function, *Ann. Math.*, **75** (1962), 47–62.

19. N. S. Mendelsohn and A. L. Dulmage, Some generalizations of the problem of distinct representatives, *Canad. Jour. Math.*, **10** (1958), 230–241.

20. O. Ore, Graphs and matching theorems, *Duke Math. Jour.*, **22** (1955), 625–639.

21. ———, *Theory of Graphs*, Amer. Math. Soc. Colloq. Publs., **38**, 1962.

22. R. Rado, Factorization of even graphs, *Quarterly Jour. Math.*, **20** (1949), 95–104.

23. H. J. Ryser, A combinatorial theorem with an application to Latin rectangles, *Proc. Amer. Math. Soc.*, **2** (1951), 550–552.

CHAPTER **6**

MATRICES OF ZEROS AND ONES

1. The class $\mathfrak{A}(\mathbf{R}, \mathbf{S})$. Let A be a $(0, 1)$-matrix of size m by n. Let the sum of row i of A be denoted by r_i and let the sum of column j of A be denoted by s_j. We call the vector

$$(1.1) \qquad R = (r_1, r_2, \ldots, r_m)$$

the *row sum vector* and the vector

$$(1.2) \qquad S = (s_1, s_2, \ldots, s_n)$$

the *column sum vector* of A. We say R is *monotone* provided $r_1 \geqq r_2 \geqq \cdots \geqq r_m$, and a similar definition holds for S. If τ denotes the total number of 1's in A, then it is clear that

$$(1.3) \qquad \tau = \sum_{i=1}^{m} r_i = \sum_{j=1}^{n} s_j.$$

The vectors R and S determine a class

$$(1.4) \qquad \mathfrak{A} = \mathfrak{A}(R, S)$$

consisting of all $(0, 1)$-matrices of size m by n with row sum vector R and column sum vector S. In this chapter we investigate the structure of the class $\mathfrak{A}$. Our theorems concern $(0, 1)$-matrices. But each of the conclusions may be rephrased in the purely combinatorial terms of set and element. This is because a (0, 1)-matrix of size m by n may be regarded as an incidence matrix for subsets T_1, T_2, ..., T_m of an n-set T.

Now let $R = (r_1, r_2, ..., r_m)$ and $S = (s_1, s_2, ..., s_n)$ be two vectors whose components are nonnegative integers. Let $\mathfrak{A} = \mathfrak{A}(R, S)$ denote the class of all (0, 1)-matrices of size m by n with row sum vector R and column sum vector S. We study the conditions under which the class $\mathfrak{A}$ is nonempty. We begin by introducing some notation. Let

(1.5) $\quad \delta_i = (1, 1, ..., 1, 0, 0, ..., 0) \quad (i = 1, 2, ..., m)$

be a vector of n components with 1's in the first r_i positions and 0's elsewhere. A matrix of the form

$$(1.6) \quad \bar{A} = \begin{bmatrix} \delta_1 \\ \delta_2 \\ \cdot \\ \cdot \\ \cdot \\ \delta_m \end{bmatrix}$$

is called *maximal*, and we call $\bar{A}$ the *maximal matrix* with row sum vector R. The column sum vector $\bar{S} = (\bar{s}_1, \bar{s}_2, ..., \bar{s}_n)$ of $\bar{A}$ is monotone. Also,

$$(1.7) \quad \sum_{i=1}^{m} r_i = \sum_{j=1}^{n} \bar{s}_j$$

and the class $\mathfrak{A}(R, \bar{S})$ contains only one matrix, namely $\bar{A}$. Let $S = (s_1, s_2, ..., s_n)$ and $S^* = (s_1^*, s_2^*, ..., s_n^*)$ be two vectors whose components are nonnegative integers. The vector S is *majorized* by S^*, written

(1.8) $\quad\quad\quad\quad S \prec S^*,$

provided that with subscripts renumbered

Sec. 1   THE CLASS $\mathfrak{A}(R, S)$ 63

(1.9) $s_1 \geqq s_2 \geqq \cdots \geqq s_n, \qquad s_1^* \geqq s_2^* \geqq \cdots \geqq s_n^*,$

(1.10) $s_1 + s_2 + \cdots + s_i \leqq s_1^* + s_2^* + \cdots + s_i^*$
$$(i = 1, 2, \ldots, n-1),$$

(1.11) $s_1 + s_2 + \cdots + s_n = s_1^* + s_2^* + \cdots + s_n^*.$

THEOREM 1.1. *Let $R = (r_1, r_2, \ldots, r_m)$ and $S = (s_1, s_2, \ldots, s_n)$ be two vectors whose components are nonnegative integers, and let $\bar{A}$ be the m by n maximal matrix with row sum vector R and column sum vector $\bar{S}$. Then the class $\mathfrak{A}(R, S)$ is nonempty if and only if*

(1.12) $\qquad\qquad S \prec \bar{S}.$

Proof. Let the class $\mathfrak{A}$ contain a matrix A. Then A may be constructed from $\bar{A}$ by shifting 1's in the rows of $\bar{A}$. Thus for S monotone we may write

(1.13) $s_1 + s_2 + \cdots + s_i \leqq \bar{s}_1 + \bar{s}_2 + \cdots + \bar{s}_i$
$$(i = 1, 2, \ldots, n-1),$$

(1.14) $s_1 + s_2 + \cdots + s_n = \bar{s}_1 + \bar{s}_2 + \cdots + \bar{s}_n,$

whence $S \prec \bar{S}$.

Now suppose that $S \prec \bar{S}$. We renumber subscripts so that R and S are monotone and construct a specific matrix $\tilde{A}$ with row sum vector R and column sum vector S. We construct $\tilde{A}$ from $\bar{A}$ by successively shifting 1's from left to right in the rows of $\bar{A}$. We first describe the construction and then verify that it may be carried out. The construction begins by shifting the last 1 in certain rows of $\bar{A}$ to column n. We require column n to have sum s_n, and the 1's are to appear in those rows in which $\bar{A}$ has the s_n largest row sums. In the event that certain row sums have equal values, preference is given to the bottommost positions. This gives us a matrix of the form

(1.15) $\qquad\qquad [\bar{A}_{n-1}, A_1].$

Here $\bar{A}_{n-1}$ is an m by $n-1$ maximal matrix with a monotone row sum vector and a monotone column sum vector. The matrix A_1 is an m by 1 matrix with column sum s_n. We now leave the matrix A_1 unaltered and iterate this construction on $\bar{A}_{n-1}$. At stage $n - f$ of the construction we have a matrix of the form

(1.16) $\qquad [\bar{A}_f, A_{n-f}].$

Here $\bar{A}_f$ is an m by f maximal matrix with a monotone row sum vector and a monotone column sum vector. The matrix A_{n-f} is of size m by $n - f$ and has monotone column sum vector $(s_{f+1}, s_{f+2}, \ldots, s_n)$.

We now proceed to stage $n - f + 1$ of the construction and verify that the construction may be carried out. Suppose that we attempt to transform the $\bar{A}_f$ of (1.16) by the prescribed procedure and are unable to attain for column f the column sum s_f. Let the monotone column sum vector of $\bar{A}_f$ be $(e_1, e_2, \ldots, e_f)$. Then we must have $e_1 < s_f$ or $e_f > s_f$. If $e_1 < s_f$, then

(1.17) $\quad s_1 + s_2 + \cdots + s_f = e_1 + e_2 + \cdots + e_f$

$$\leq f e_1 < f s_f \leq s_1 + s_2 + \cdots + s_f,$$

and this is a contradiction. On the other hand, if $e_f > s_f$, then $e_f > s_{f+1}, s_{f+2}, \ldots, s_n$ so that the first e_f rows of A_{n-f} contain at least one 0 in each column. The matrix $\bar{A}_f$ is maximal with a monotone row sum vector. But then by the construction the last $m - e_{f-1}$ rows of A_{n-f} contain only 0's. Hence we have

(1.18) $\quad e_1 + e_2 + \cdots + e_{f-1} = \bar{s}_1 + \bar{s}_2 + \cdots + \bar{s}_{f-1}.$

By (1.12) and (1.18),

(1.19)

$$s_1 + s_2 + \cdots + s_{f-1} + s_f = e_1 + e_2 + \cdots + e_{f-1} + e_f$$
$$\leq \bar{s}_1 + \bar{s}_2 + \cdots + \bar{s}_{f-1} + s_f$$
$$= e_1 + e_2 + \cdots + e_{f-1} + s_f.$$

But then $s_f \geqq e_f$ and this contradicts $e_f > s_f$. We may begin the construction with column n, and the process comes to an automatic termination with column 1. This proves Theorem 1.1. Note that if R_i denotes the row sum vector of the first i columns of $\tilde{A}$, then R_i is monotone for each $i = 1, 2, \ldots, n$.

The matrix

$$(1.20) \qquad \tilde{A} = \begin{bmatrix} 1 & 1 & 1 & 0 \\ 1 & 1 & 0 & 1 \\ 1 & 0 & 0 & 0 \\ 0 & 1 & 0 & 0 \end{bmatrix}$$

illustrates the construction for $R = S = (3, 3, 1, 1)$.

In this section we have determined the conditions under which the class $\mathfrak{A}$ is nonempty. A much more difficult problem is the determination of the precise number of matrices in the class. This number is undoubtedly an extremely intricate function of R and S.

2. An application to Latin rectangles. Let there be given an r by s Latin rectangle based on n elements labeled $1, 2, \ldots, n$. In this section we obtain a necessary and sufficient condition for the extension of the Latin rectangle to a Latin square of order n.

THEOREM 2.1. *Let A be a $(0, 1)$-matrix of size m by n with $m \leqq n$. Let A have row sum vector $R = (k, k, \ldots, k)$, where k is a positive integer. Let A have column sum vector $S = (s_1, s_2, \ldots, s_n)$, where*

$$(2.1) \qquad 0 \leqq k - s_i \leqq n - m \qquad (i = 1, 2, \ldots, n).$$

Then

$$(2.2) \qquad A = P_1 + P_2 + \cdots + P_k,$$

where the P_i are permutation matrices.

Proof. We may assume that $m < n$. For if $m = n$, then the theorem reduces to Theorem 5.3 of Chapter 5. Now there exists a $(0, 1)$-matrix A' of size $n - m$ by n with row sum vector $R' = (k, k, \ldots, k)$ and column sum vector $S' = (k - s_1, k - s_2, \ldots, k - s_n)$. For let $\bar{S}'$ be the vector with the first k components equal to $n - m$ and the remaining $n - k$ components equal to 0. Then by (2.1), $S' \prec \bar{S}'$. Hence the preceding theorem establishes the existence of A'. But then

$$(2.3) \qquad \begin{bmatrix} A \\ A' \end{bmatrix}$$

is an n by n matrix with each row and column sum equal to k. Hence this matrix is a sum of k permutation matrices. But then A is a sum of k permutation matrices.

THEOREM 2.2. *Let there be given an r by s Latin rectangle based on n elements labeled $1, 2, \ldots, n$. Let $N(i)$ denote the number of times that i occurs in the Latin rectangle. The Latin rectangle may be extended to a Latin square of order n if and only if*

$$(2.4) \qquad N(i) \geqq r + s - n \qquad (i = 1, 2, \ldots, n).$$

Proof. Let T be the n-set of elements $1, 2, \ldots, n$ and let T_j be the set of all elements of T that do not appear in row j of the Latin rectangle ($j = 1, 2, \ldots, r$). Then each T_j is an $(n - s)$-subset of T. Let $M(i)$ denote the number of times that i appears among the sets $T_1, T_2, \ldots, T_r$ ($i = 1, 2, \ldots, n$). Now if the Latin rectangle may be extended to an n by n Latin square, then $M(i) \leqq n - s$. But $N(i) + M(i) = r$, whence $N(i) \geqq r + s - n$.

Conversely, suppose that $N(i) \geqq r + s - n$. Let A be the incidence matrix for the subsets $T_1, T_2, \ldots, T_r$ of the n-set T. The r by n matrix A has row sum vector $R = (n - s, n - s, \ldots, n - s)$ and column sum vector $S = (M(1), M(2), \ldots, M(n))$. By hypothesis $N(i) = r - M(i)$

$\geq r + s - n$, and since our configuration is a Latin rectangle, $N(i) = r - M(i) \leq s$. Hence

(2.5) $\quad 0 \leq n - s - M(i) \leq n - r \quad (i = 1, 2, \ldots, n)$,

and by Theorem 2.1,

(2.6) $\quad A = P_1 + P_2 + \cdots + P_{n-s}$,

where the P_i are permutation matrices. But these permutation matrices define r-permutations of $1, 2, \ldots, n$, and these r-permutations may be adjoined as columns to the r by s Latin rectangle to yield an r by n Latin rectangle. Then this configuration may be extended to an n by n Latin square by Theorem 3.1 of Chapter 5. Or we may avoid the use of this theorem and complete the transposed n by r Latin rectangle to an n by n Latin square by the method just described. The condition on $N(i)$ is trivially satisfied in this case.

The preceding theorem suggests the following problem. Let there be given an n by n array based on the elements $1, 2, \ldots, n$ and an unassigned element x. Under what conditions does a suitable replacement of the x's by $1, 2, \ldots, n$ yield a Latin square of order n? The theorem solves this problem in a very special case. The general problem has never been successfully handled.

3. Interchanges. Let A be a matrix in the class $\mathfrak{A} = \mathfrak{A}(R, S)$ of all $(0, 1)$-matrices of size m by n with row sum vector $R = (r_1, r_2, \ldots, r_m)$ and column sum vector $S = (s_1, s_2, \ldots, s_n)$. Consider the 2 by 2 submatrices of A of the types

(3.1) $\quad A_1 = \begin{bmatrix} 1 & 0 \\ 0 & 1 \end{bmatrix}, \quad A_2 = \begin{bmatrix} 0 & 1 \\ 1 & 0 \end{bmatrix}.$

An *interchange* is a transformation of the elements of A that changes a submatrix of type A_1 into type A_2 or vice versa and leaves all other elements of A unaltered. This is, in a sense, the most elementary operation that may be applied to A to yield a new matrix in the class $\mathfrak{A}$. Interchanges are helpful in dealing with many problems involving the class $\mathfrak{A}$. We begin by establishing the interchange theorem.

THEOREM 3.1. *Let A and A' belong to $\mathfrak{A}(R, S)$. Then A is transformable into A' by a finite sequence of interchanges.*

Proof. We take R and S monotone and we let $\tilde{A}$ be the matrix constructed in § 1. We may apply interchanges to A and replace column n of A by column n of $\tilde{A}$. This is because column n of $\tilde{A}$ has its 1's in the rows with the s_n largest row sums. We now leave column n of the transformed matrix unaltered and concentrate our attention on column $n - 1$. The first $n - 1$ columns of both $\tilde{A}$ and the transformed matrix are matrices with the same row sum vector. By the structure of $\tilde{A}$ we may apply interchanges to the transformed matrix and replace column $n - 1$ of the transformed matrix by column $n - 1$ of $\tilde{A}$. In this way we transform A into $\tilde{A}$ by interchanges. We may also transform A' into $\tilde{A}$ by interchanges. Let the intermediate matrices taking A' into $\tilde{A}$ be $A_1, A_2, \ldots, A_q$. But then there exists an interchange taking $\tilde{A}$ into A_q. Also, there exists an interchange taking A_q into A_{q-1} and so on. Thus $\tilde{A}$ is transformable into A' by interchanges. Hence A is transformable into A' by interchanges. We remark that the minimal number of interchanges required to transform A into A' is apparently a hopelessly complicated function of A and A'.

Let A be a matrix in the class $\mathfrak{A}(R, S)$. In many investigations one assumes without loss of generality that the row sum vector R and the column sum vector S of A satisfy

(3.2) $$r_1 \geqq r_2 \geqq \cdots \geqq r_m > 0,$$

(3.3) $$s_1 \geqq s_2 \geqq \cdots \geqq s_n > 0.$$

This means that we have excluded rows and columns of zeros and permuted rows and columns so that they are monotone. A nonempty class $\mathfrak{A}(R, S)$ with R and S satisfying (3.2) and (3.3) is called *normalized*. Henceforth throughout our discussion we take $\mathfrak{A}$ normalized.

Let A be a matrix in the normalized class $\mathfrak{A}$. An element $a_{ef} = 1$ of A is an *invariant* 1 provided that no sequence of interchanges applied to A replaces $a_{ef} = 1$ by 0. If $a_{ef} = 1$ is an invariant 1 of A, then by the interchange theorem the entries in the (e, f) position of all of the matrices in $\mathfrak{A}$ must be invariant 1's. Thus all or none of the matrices in $\mathfrak{A}$ contain invariant 1's, and we say $\mathfrak{A}$ is with or without invariant 1's.

THEOREM 3.2. *The normalized class $\mathfrak{A}$ is with invariant 1's if and only if every matrix in $\mathfrak{A}$ has a decomposition of the form*

$$(3.4) \qquad A = \begin{bmatrix} J & * \\ * & 0 \end{bmatrix}.$$

Here J is a matrix of 1's of size e by f $(0 < e \leq m; 0 < f \leq n)$ and 0 is a zero matrix. The integers e and f are not necessarily unique, but they are determined by the row sum vector R and the column sum vector S and are independent of the particular choice of A in $\mathfrak{A}$.

Proof. Clearly every 1 in J of (3.4) is an invariant 1. Suppose then that the normalized class $\mathfrak{A}$ is with invariant 1's. Let a_{ef} be an invariant 1 with $e + f$ maximal and let

$$(3.5) \qquad A = \begin{bmatrix} W & X \\ Y & Z \end{bmatrix},$$

where W is of size e by f. If a 0 appears in W, then by the normalization of $\mathfrak{A}$ two interchanges at most are required to replace the invariant 1 by 0. Hence $W = J$ and each of the 1's in J is an invariant 1. Now since $e + f$ is maximal, we may select A so that a 0 appears in column 1 of X.

Now if Z contains a 1 in row t, then we may apply an interchange if necessary and assume that column 1 of Z contains a 1 in row t. But then Y contains only 1's in row t. For if Y contains a 0 in row t, then an interchange moves an invariant 1 of A. In fact each of the 1's in row t of Y is an invariant 1. But this contradicts $e + f$ maximal. Hence $Z = 0$ and A is of the form (3.4). But then every A in $\mathfrak{A}$ is of the form (3.4).

4. Maximal term rank. Let $\check{\rho}$ be the minimal term rank and let $\bar{\rho}$ be the maximal term rank of the matrices in the normalized class $\mathfrak{A} = \mathfrak{A}(R, S)$. We devote this section to an analysis of $\bar{\rho}$. The theorems described are of interest in their own right and yield an explicit formula for $\bar{\rho}$. But the techniques employed in the derivations of these theorems are applicable to a number of related problems. We begin with an elementary result on intermediate term ranks.

THEOREM 4.1. *Let $\check{\rho}$ be the minimal term rank and let $\bar{\rho}$ be the maximal term rank of the matrices in the normalized class $\mathfrak{A}$. Then $\mathfrak{A}$ contains a matrix A_ρ of term rank ρ, where ρ is an arbitrary integer in the interval*

(4.1) $$\check{\rho} \leq \rho \leq \bar{\rho}.$$

Proof. An interchange applied to a matrix A in $\mathfrak{A}$ changes the term rank by 1 or leaves the term rank unaltered. But by the interchange theorem we may transform a matrix $A_{\bar{\rho}}$ of term rank $\bar{\rho}$ into a matrix $A_{\check{\rho}}$ of term rank $\check{\rho}$. This implies that there exists a matrix A_ρ of term rank ρ.

THEOREM 4.2. *The normalized class $\mathfrak{A}$ contains a matrix $A_{\bar{\rho}}$ with $\bar{\rho}$ 1's in the positions $(1, \bar{\rho})$, $(2, \bar{\rho} - 1)$, ..., $(\bar{\rho}, 1)$.*

Proof. Let $A_{\bar{\rho}}$ be a matrix of maximal term rank $\bar{\rho}$. We select a specified set of $\bar{\rho}$ 1's of $A_{\bar{\rho}}$ with no two on a line, and we call these 1's the *essential* 1's of $A_{\bar{\rho}}$. All other 1's of $A_{\bar{\rho}}$ we call *unessential*. We may select an $A_{\bar{\rho}}$ with the $\bar{\rho}$ essential 1's in the first $\bar{\rho}$ rows. For suppose that an essential 1 is in the (i, j) position and row i' contains no essential 1 ($i' \leq \bar{\rho} < i$). Then if a 1 is in the (i', j) position, we call this 1 essential and the 1 in the (i, j) position unessential. On the other hand, if a 0 is in the (i', j) position, then by the normalization of $\mathfrak{A}$ there is an interchange that places the essential 1 in the (i', j) position and retains term rank $\bar{\rho}$. In this way we obtain an $A_{\bar{\rho}}$ with the essential 1's in the first $\bar{\rho}$ rows. A similar argument applied to the columns yields a matrix of the form

$$(4.2) \qquad A_{\bar{\rho}} = \begin{bmatrix} D & * \\ * & 0 \end{bmatrix}.$$

Here D is of order and term rank $\bar{\rho}$. The matrix 0 is a zero matrix because the term rank of $A_{\bar{\rho}}$ cannot exceed $\bar{\rho}$. We say that the $\bar{\rho}$ entries in the positions $(1, \bar{\rho})$, $(2, \bar{\rho} - 1)$, ..., $(\bar{\rho}, 1)$ comprise the *secondary diagonal* of D. We now obtain an $A_{\bar{\rho}}$ with $\bar{\rho}$ essential 1's on the secondary diagonal of D. Suppose that essential 1's are in the $(1, \bar{\rho})$, $(2, \bar{\rho} - 1)$, ..., $(d, \bar{\rho} - d + 1)$ positions of D but that there is no essential 1 in the $(d + 1, \bar{\rho} - d)$ position of D. Then the essential 1 in row $d + 1$ and the essential 1 in column $\bar{\rho} - d$ determine a 2 by 2 submatrix of D of one of the following four types:

$$(4.3) \quad \begin{bmatrix} 1 & 0 \\ 0 & 1 \end{bmatrix}, \quad \begin{bmatrix} 1 & 0 \\ 1 & 1 \end{bmatrix}, \quad \begin{bmatrix} 1 & 1 \\ 0 & 1 \end{bmatrix}, \quad \begin{bmatrix} 1 & 1 \\ 1 & 1 \end{bmatrix}.$$

The 1's on the main diagonal in (4.3) correspond to essential 1's of $A_{\bar{\rho}}$. In each case at most one interchange of $A_{\bar{\rho}}$ places two 1's in positions corresponding to the secondary diagonal in (4.3). These 1's may be regarded as essential 1's of $A_{\bar{\rho}}$. This gives us an essential 1 in the $(d + 1,$

$\bar{\rho} - d$) position of D. Thus there is an $A_{\bar{\rho}}$ with $\bar{\rho}$ essential 1's on the secondary diagonal of D.

Let Q be a (0, 1)-matrix. We let $N_0(Q)$ denote the number of 0's in Q and we let $N_1(Q)$ denote the number of 1's in Q. Let A be a (0, 1)-matrix of size m by n. Then A is required to have $m, n > 0$. But if A is partitioned into blocks, it is convenient to allow degenerate submatrices W of size e by f with e or f equal to 0. A degenerate submatrix W has $N_0(W) = N_1(W) = 0$. We are now ready to state our main conclusion concerning the maximal term rank $\bar{\rho}$ of the matrices in the normalized class $\mathfrak{A}$.

THEOREM 4.3. *Every matrix A in the normalized class $\mathfrak{A}$ has a decomposition of the form*

$$(4.4) \qquad A = \begin{bmatrix} W & X \\ Y & Z \end{bmatrix}.$$

Here W is of size e by f ($0 \leq e \leq m$; $0 \leq f \leq n$), Z is of size $m - e$ by $n - f$, and

$$(4.5) \qquad N_0(W) + N_1(Z) = \bar{\rho} - (e + f).$$

The integers e and f are not necessarily unique, but they are determined by the row sum vector R and the column sum vector S and are independent of the particular choice of A in $\mathfrak{A}$. In particular, the matrix $A_{\bar{\rho}}$ of Theorem 4.2 satisfies

$$(4.6) \qquad N_0(W) = 0, \qquad N_1(Z) = \bar{\rho} - (e + f).$$

Proof. If A has row sum vector $R = (r_1, r_2, \ldots, r_m)$ and column sum vector $S = (s_1, s_2, \ldots, s_n)$, it follows at once that

$$(4.7) \quad N_0(W) + N_1(Z) = ef + (r_{e+1} + r_{e+2} + \cdots + r_m) \\ - (s_1 + s_2 + \cdots + s_f).$$

Thus $N_0(W) + N_1(Z)$ is independent of the particular choice of A in $\mathfrak{A}$. This means that it suffices to prove the theorem for the matrix $A_{\bar{p}}$ of Theorem 4.2. We confine our attention to $A_{\bar{p}}$ and use induction on the rows of $A_{\bar{p}}$. The theorem is valid for matrices of one row. The induction hypothesis asserts the statement of the theorem for matrices of $m - 1$ rows, and we prove the theorem for matrices of m rows. Now if $\bar{p} = m$, the theorem is valid with $e = m$ and $f = 0$. Also, if $\bar{p} = n$, the theorem is valid with $e = 0$ and $f = n$. Hence we assume that $\bar{p} < m, n$. Note that in this case the e and the f of the theorem yield a proper decomposition of $A_{\bar{p}}$ ($0 < e < m$; $0 < f < n$). For if $e = 0$ or m or if $f = 0$ or n, then we contradict $\bar{p} < m, n$.

Now let $A_{\bar{p}} = [a_{ij}]$ of Theorem 4.2 have term rank $\bar{p} < m, n$. We normalize the first row of $A_{\bar{p}}$ by interchanges in the following way. If $s_i > s_j$ and $a_{1i} = 0$ and $a_{1j} = 1$, we apply an interchange that replaces a_{1i} by 1 and a_{1j} by 0. If $i < \bar{p}$, then the interchange is selected so that it does not move the essential 1 in the $(\bar{p} - i + 1, i)$ position of $A_{\bar{p}}$. Also, if $s_i = s_j$ ($i < j$) and $a_{1i} = 1$ and $a_{1j} = 0$, we apply an interchange that replaces a_{1i} by 0 and a_{1j} by 1. If $j < \bar{p}$, then the interchange is not to move the essential 1 in the $(\bar{p} - j + 1, j)$ position of $A_{\bar{p}}$. If this cannot be avoided, then the matrix has a 1 in the $(\bar{p} - i + 1, j)$ position, and a second interchange involving rows $\bar{p} - i + 1$ and $\bar{p} - j + 1$ restores a 1 to the $(\bar{p} - j + 1, j)$ position. If we apply all possible interchanges of the type described, we obtain a matrix of the following form:

$$(4.8) \qquad M = \begin{bmatrix} \delta_1 & \delta_2 \\ A_{\bar{p}-1} & 0 \end{bmatrix}.$$

In (4.8) δ_1 is a vector of n' components and δ_2 is a vector of $n - n'$ 1's. The degenerate case $n = n'$ is not excluded. The matrix $A_{\bar{p}-1}$ of size $m - 1$ by n' has 1's in the $(1, \bar{p} - 1)$, $(2, \bar{p} - 2)$, ..., $(\bar{p} - 1, 1)$ positions and generates a normalized class $\mathfrak{A}'$. The matrix 0 is a zero matrix of size $m - 1$ by $n - n'$.

We prove that $\bar{p} - 1$ is the maximal term rank of the matrices in $\mathfrak{A}'$. Let A' be a matrix in $\mathfrak{A}'$ with $\bar{p}$ essential 1's in the $(1, \bar{p})$, $(2, \bar{p} - 1)$, ..., $(\bar{p}, 1)$ positions. In (4.8) we replace $A_{\bar{p}-1}$ by A' and call the resulting matrix M'. The matrix M' belongs to the class $\mathfrak{A}$. If needed, an interchange is available to place a 1 in the $(1, n)$ position of M'. But then the term rank of M' contradicts the maximality of $\bar{p}$ for $\mathfrak{A}$. Thus $A_{\bar{p}-1}$ is of maximal term rank $\bar{p} - 1$ with essential 1's in the $(1, \bar{p} - 1)$, $(2, \bar{p} - 2)$, ..., $(\bar{p} - 1, 1)$ positions. By the induction hypothesis

$$(4.9) \qquad A_{\bar{p}-1} = \begin{bmatrix} W' & X' \\ Y' & Z' \end{bmatrix},$$

where W' is a matrix of 1's of size e' by f' and $N_1(Z') = \bar{p} - 1 - (e' + f')$. We select f' maximal in the sense that each of the columns of X' above the essential 1's in Z' contains a 0. We know that $\bar{p} - 1 < m - 1$, but we may have $\bar{p} - 1 = n'$. The latter leads to a degenerate decomposition of $A_{\bar{p}-1}$ with $e' = 0$ and $f' = n'$. But all other decompositions have $0 < e' < m - 1$ and $0 < f' < n'$.

Now in (4.8) suppose that M has a 0 of δ_1 above a column of Y' and let this column of Y' contain an unessential 1 of $A_{\bar{p}-1}$. Then by our normalization of the first row of M, we know that $n' = n$ and that 0's appear in positions $f' + 1, f' + 2, \ldots, n$ of δ_1. Now if the decomposition of $A_{\bar{p}-1}$ is degenerate, this contradicts $\bar{p} - 1 = n'$, and if the decomposition of $A_{\bar{p}-1}$ is nondegenerate, this contradicts $r_1 \geq r_2$. Hence if M has a 0 of δ_1 above a column of Y', then the column of Y' contains no unessential 1 of $A_{\bar{p}-1}$. This means that the decomposition (4.9) may be adjusted so that M has no 0's in δ_1 above the columns of Y'. If needed, an interchange is available to place a 1 in the $(1, \bar{p})$ position of M. This gives us a decomposition of the type desired.

An easy consequence of this theorem is a remarkable

formula for $\bar{\rho}$ in terms of the components of the row sum vector $R = (r_1, r_2, \ldots, r_m)$ and the column sum vector $S = (s_1, s_2, \ldots, s_n)$ of the matrices in the normalized class $\mathfrak{A}(R, S)$.

THEOREM 4.4. *Let*

(4.10)
$$t_{ij} = ij + (r_{i+1} + r_{i+2} + \cdots + r_m) - (s_1 + s_2 + \cdots + s_j)$$
$$(i = 0, 1, \ldots, m; j = 0, 1, \ldots, n).$$

Then

(4.11) $$\bar{\rho} = \min_{i,j} \{t_{ij} + (i + j)\}$$
$$(i = 0, 1, \ldots, m; j = 0, 1, \ldots, n).$$

Proof. Let

(4.12) $$A_{\bar{\rho}} = \begin{bmatrix} W & X \\ Y & Z \end{bmatrix}$$

be a matrix of maximal term rank $\bar{\rho}$ and let W be of size i by j. Now $\bar{\rho}$ lines suffice to cover the 1's in $A_{\bar{\rho}}$. Hence

(4.13) $$N_1(Z) + (i + j) \geq \bar{\rho}.$$

But $N_0(W) \geq 0$ so that

(4.14) $$t_{ij} + (i + j) \geq \bar{\rho}.$$

Equality is attained in (4.14) for the e and the f of Theorem 4.3. This establishes Theorem 4.4.

A formula is also available for the evaluation of $\tilde{\rho}$, but we do not pursue this topic here. The following theorem gives conditions that insure $\tilde{\rho} < \bar{\rho}$.

THEOREM 4.5. *Let the normalized class* $\mathfrak{A}$ *be without invariant* 1's *and let* $\bar{\rho} < m, n$. *Then* $\tilde{\rho} < \bar{\rho}$.

Proof. By hypothesis $\bar{\rho} < m, n$. Hence the e and the f of Theorem 4.3 are in the intervals $0 < e < m$ and $0 <$

$f < n$. By hypothesis the entry in the $(1, 1)$ position of the matrix $A_{\bar{\rho}}$ of Theorem 4.3 is not an invariant 1. But in Theorem 4.3 we have $N_0(W) + N_1(Z) = \bar{\rho} - (e + f)$. This means that there exist matrices in $\mathfrak{A}$ with fewer than $\bar{\rho} - (e + f)$ 1's in Z. But then the 1's in such a matrix may be covered by fewer than $\bar{\rho}$ lines. Hence $\tilde{\rho} < \bar{\rho}$.

Note that Theorem 4.5 need not be valid if the hypothesis on invariant 1's is deleted. The class consisting of a maximal matrix $\bar{A}$ has $\tilde{\rho} = \bar{\rho}$. Nor may the restriction $\bar{\rho} < m, n$ be removed. For example, the class of the $n!$ permutation matrices of order n has $\tilde{\rho} = \bar{\rho}$. An unsettled problem of interest calls for a neat classification of all classes $\mathfrak{A}(R, S)$ with $\tilde{\rho} = \bar{\rho}$.

5. Related problems.

The preceding section has been devoted to an analysis of the term rank of the matrices A in the normalized class $\mathfrak{A}$. We may associate other worthwhile functions with A and investigate their behavior as A ranges over its class. Such investigations have been carried out in certain instances. For example, let $\tilde{\sigma}$ be the minimal trace and let $\bar{\sigma}$ be the maximal trace of the matrices in the normalized class $\mathfrak{A}$. Then one may prove that

(5.1) $$\tilde{\sigma} = \max_{i,j} \{\min(i, j) - t_{ij}\}$$

$$(i = 0, 1, \ldots, m; j = 0, 1, \ldots, n)$$

and

(5.2) $$\bar{\sigma} = \min_{i,j} \{t_{ij} + \max(i, j)\}$$

$$(i = 0, 1, \ldots, m; j = 0, 1, \ldots, n).$$

These formulas bear a striking resemblance to the formula for $\bar{\rho}$ in § 4, and their derivations proceed along rather similar lines.

But many extremal problems of this variety cannot be disposed of with such thoroughness. Let A be in the normalized class $\mathfrak{A}(R, S)$ and let α be an integer in the interval $1 \leq \alpha \leq r_m$. Let E be an m by ϵ submatrix of A and let each row sum of E be at least α. The minimal ϵ with this property is called the α-*width* $\epsilon(\alpha)$ of A. The integer α and the matrix A uniquely determine $\epsilon(\alpha)$. Now let $\tilde{\epsilon}(\alpha)$ be the minimal α-width and let $\bar{\epsilon}(\alpha)$ be the maximal α-width of the matrices in $\mathfrak{A}$. One may prove that the matrix $\tilde{A}$ constructed in § 1 has α-width $\tilde{\epsilon}(\alpha)$ for each $\alpha = 1, 2, \ldots, r_m$. In fact, it turns out that the αth 1 in the last row of $\tilde{A}$ occurs in column $\tilde{\epsilon}(\alpha)$. This remarkable behavior of $\tilde{A}$ gives an efficient procedure for the evaluation of $\tilde{\epsilon}(\alpha)$. But very little is known about the behavior of $\bar{\epsilon}(\alpha)$. More information here would be of the utmost value. This will become apparent in Chapter 8. There we mention an interconnection between $\bar{\epsilon}(1)$ and finite projective planes.

Certain special classes possess intriguing problems in their own right. Let $\mathfrak{A}(K, K)$ denote the class defined by $m = n$ and

(5.3) $$R = S = K = (k, k, \ldots, k),$$

where k is a fixed integer in the interval $1 \leq k \leq n$. Thus $\mathfrak{A}(K, K)$ is the class of all $(0, 1)$-matrices of order n with exactly k 1's in each row and column. For $k = 1$ the class consists of the $n!$ permutation matrices of order n, and for $k = n$ the class consists of the matrix J of order n. Now by Theorem 5.3 of Chapter 5, the class $\mathfrak{A}(K, K)$ has

(5.4) $$\tilde{\rho} = \bar{\rho} = m = n.$$

Under these conditions it is natural to ask for the minimal permanent and the maximal permanent of the matrices in $\mathfrak{A}(K, K)$. But neither value has been determined. The minimal value may have considerable combinatorial significance. The corresponding problem for doubly stochastic matrices leads to the van der Waerden conjecture of Chapter 5.

References for Chapter 6

Theorem 1.1 is by Gale [9] and Ryser [13]. Theorem 4.2 is by Haber [10]. The remaining theorems in § 2, § 3, and § 4 are by Ryser [12; 13; 14]. The proofs in § 3 follow Haber [10]. The minimal term rank $\bar{\rho}$ is discussed in Haber [10; 11]. Traces are investigated by Fulkerson [5] and Ryser [15] and α-widths by Fulkerson and Ryser [6; 7; 8].

1. A. L. Dulmage and N. S. Mendelsohn, Coverings of bipartite graphs, *Canad. Jour. Math.*, **10** (1958), 517–534.

2. ———, The term and stochastic ranks of a matrix, *Canad. Jour. Math.*, **11** (1959), 269–279.

3. T. Evans, Embedding incomplete Latin squares, *Amer. Math. Monthly*, **67** (1960), 958–961.

4. L. R. Ford, Jr., and D. R. Fulkerson, *Flows in Networks*, Princeton University Press, 1962.

5. D. R. Fulkerson, Zero-one matrices with zero trace, *Pacific Jour. Math.*, **10** (1960), 831–836.

6. D. R. Fulkerson and H. J. Ryser, Widths and heights of (0, 1)-matrices, *Canad. Jour. Math.*, **13** (1961), 239–255.

7. ———, Multiplicities and minimal widths for (0, 1)-matrices, *Canad. Jour. Math.*, **14** (1962), 498–508.

8. ———, Width sequences for special classes of (0, 1)-matrices, *Canad. Jour. Math.*, **15** (1963), 371–396.

9. D. Gale, A theorem on flows in networks, *Pacific Jour. Math.*, **7** (1957), 1073–1082.

10. R. M. Haber, Term rank of (0, 1)-matrices, *Rend. Sem. Mat. Padova*, **30** (1960), 24–51.

11. ———, Minimal term rank of a class of (0, 1)-matrices, *Canad. Jour. Math.*, **15** (1963), 188–192.

12. H. J. Ryser, A combinatorial theorem with an application to Latin rectangles, *Proc. Amer. Math. Soc.*, **2** (1951), 550–552.

13. ———, Combinatorial properties of matrices of zeros and ones, *Canad. Jour. Math.*, **9** (1957), 371–377.

14. ———, The term rank of a matrix, *Canad. Jour. Math.*, **10** (1958), 57–65.

15. ———, Traces of matrices of zeros and ones, *Canad. Jour. Math.*, **12** (1960), 463–476.

16. ———, Matrices of zeros and ones, *Bull. Amer. Math. Soc.*, **66** (1960), 442–464.

CHAPTER 7

ORTHOGONAL LATIN SQUARES

1. Existence theorems. Let $A_1 = [a_{ij}^{(1)}]$ and $A_2 = [a_{ij}^{(2)}]$ denote two n by n Latin squares based on n elements labeled 1, 2, ..., n and let $n \geq 3$. The Latin squares A_1 and A_2 are called *orthogonal* provided that the n^2 2-samples

$$(1.1) \qquad (a_{ij}^{(1)}, a_{ij}^{(2)}) \qquad (i, j = 1, 2, \ldots, n)$$

are distinct. In other words suppose that one of the Latin squares is superimposed on the other. Then the resulting configuration is an n by n array of ordered pairs of 1, 2, ..., n, and the orthogonality requirement on the Latin squares means that all of the entries in this array are distinct. The n by n array of 2-samples formed from a pair of orthogonal Latin squares is often referred to in the literature as a *Graeco-Latin square* or an *Euler square*. The following

$$(1.2) \qquad A_1 = \begin{bmatrix} 1 & 2 & 3 \\ 2 & 3 & 1 \\ 3 & 1 & 2 \end{bmatrix}, \quad A_2 = \begin{bmatrix} 1 & 2 & 3 \\ 3 & 1 & 2 \\ 2 & 3 & 1 \end{bmatrix}$$

is an illustration of a pair of orthogonal Latin squares of order 3.

More generally, let $A_1, A_2, \ldots, A_t$ be a set of two or more Latin squares of order $n \geq 3$. The Latin squares in this set are called *orthogonal*, and we refer to $A_1, A_2, \ldots, A_t$ as an *orthogonal set*, provided that A_i and A_j are orthogonal for each $i \neq j$. In this chapter we study orthogonal sets. These configurations have a long history and were investigated at one time mainly for their recreational appeal. But today we recognize that they are important in the study of finite projective planes and related topics. We begin with the following elementary result.

THEOREM 1.1. *Let $A_1, A_2, \ldots, A_t$ be a set of t orthogonal Latin squares of order $n \geq 3$. Then*

$$(1.3) \qquad t \leq n - 1.$$

Proof. We relabel the elements of each of the Latin squares so that the first row of each of the Latin squares consists of the elements $1, 2, \ldots, n$ in this order. This does not destroy the orthogonality of the set. Consider now the t entries that appear in the $(2, 1)$ positions of the Latin squares. These t entries must be distinct, for otherwise we contradict the orthogonality of the set. Nor can one of these entries equal 1. Hence $t \leq n - 1$.

If equality holds in (1.3), then the orthogonal set is said to be *complete*. The set (1.2) is complete.

At this point we presuppose an understanding of the elementary properties of finite fields. These are the fields composed of a finite number of elements. If n denotes the number of elements in the field, it is well known that we must have $n = p^\alpha$, where p is a prime and α is a positive integer. Conversely, for every prime p and for every positive integer α, there exists a field of $n = p^\alpha$ elements. Moreover, two fields on $n = p^\alpha$ elements are unique in the sense of isomorphism. The field of $n = p^\alpha$ elements is called a *Galois field* and is designated by the notation $GF(p^\alpha)$. If $\alpha = 1$ the elements of the Galois field may be selected as the complete residue system $0, 1, \ldots, p - 1$

modulo p. The field operations of addition and multiplication are then performed modulo p. In what follows we prove an existence theorem for complete sets of orthogonal Latin squares. The proof utilizes the existence of the Galois fields $GF(p^\alpha)$.

THEOREM 1.2. *Let $n = p^\alpha$, where p is a prime and α is a positive integer. Then for $n \geq 3$ there exists a complete set of $n - 1$ orthogonal Latin squares of order n.*

Proof. Let the elements of the Galois field $GF(p^\alpha)$ be denoted by $a_0 = 0, a_1 = 1, a_2, \ldots, a_{n-1}$. Define the $n-1$ matrices of order n

(1.4) $$A_e = [a_{ij}^{(e)}] \quad (i, j = 0, 1, \ldots, n-1; \\ e = 1, 2, \ldots, n-1),$$

where

(1.5) $$a_{ij}^{(e)} = a_e a_i + a_j.$$

Each A_e of (1.4) is a Latin square. For if A_e has two equal elements in the same row, there exists a j and j' such that

(1.6) $$a_e a_i + a_j = a_e a_i + a_{j'}.$$

But this implies $a_j = a_{j'}$ and $j = j'$. Similarly, if A_e has two equal elements in the same column, there exists an i and i' such that

(1.7) $$a_e a_i + a_j = a_e a_{i'} + a_j.$$

But since $a_e \neq 0$ this implies $a_i = a_{i'}$ and $i = i'$. Hence each A_e is a Latin square. Now let $1 \leq e < f \leq n - 1$. Then A_e and A_f are orthogonal. For suppose that

(1.8) $$(a_{ij}^{(e)}, a_{ij}^{(f)}) = (a_{i'j'}^{(e)}, a_{i'j'}^{(f)}).$$

Then

(1.9) $$a_e a_i + a_j = a_e a_{i'} + a_{j'},$$

(1.10) $$a_f a_i + a_j = a_f a_{i'} + a_{j'}.$$

Subtracting (1.10) from (1.9) gives

(1.11) $$a_i(a_e - a_f) = a_{i'}(a_e - a_f).$$

But since $a_e \neq a_f$ we have $a_i = a_{i'}$ and $i = i'$. But then substituting in (1.9) gives $a_j = a_{j'}$ and $j = j'$. Hence (1.4) is an orthogonal set.

THEOREM 1.3. *For $n \geq 3$ and $t \geq 2$ a set of t orthogonal Latin squares of order n is equivalent to an n^2 by $t + 2$ array*

(1.12) $$A = [a_{ij}] \qquad (i = 1, 2, \ldots, n^2;$$

$$j = 1, 2, \ldots, t + 2).$$

The entries a_{ij} of A are elements labeled $1, 2, \ldots, n$, and the rows of each n^2 by 2 subarray of A are the n^2 2-samples of $1, 2, \ldots, n$.

Proof. Let the array A be given. We permute the rows of A so that the entries in the rows of the first two columns are in natural order $(1, 1), (1, 2), \ldots, (1, n), \ldots, (n, 1), (n, 2), \ldots, (n, n)$. Then for each $e = 3, 4, \ldots, t + 2$, we define an n by n array A_e as follows. The first row of A_e consists of the first n entries of column e of A, the second row of A_e consists of the next n entries of column e of A, and so on, until finally the last row of A_e consists of the last n entries of column e of A. Then $A_3, A_4, \ldots, A_{t+2}$ is a set of t orthogonal Latin squares of order n. For the assumptions on A are such that each of the arrays is a Latin square. In fact, column 1 of A implies that A_e does not have two equal entries in a row, and column 2 of A implies that A_e does not have two equal entries in a column. Also, if $e \neq f$, then A_e and A_f are orthogonal because of the structure of columns e and f of A. The converse proposition is proved similarly.

The following array is the one associated with the Latin squares of (1.2):

$$(1.13) \quad \begin{bmatrix} 1 & 1 & 1 & 1 \\ 1 & 2 & 2 & 2 \\ 1 & 3 & 3 & 3 \\ 2 & 1 & 2 & 3 \\ 2 & 2 & 3 & 1 \\ 2 & 3 & 1 & 2 \\ 3 & 1 & 3 & 2 \\ 3 & 2 & 1 & 3 \\ 3 & 3 & 2 & 1 \end{bmatrix}.$$

THEOREM 1.4. *If there exists a set of t orthogonal Latin squares of order n and if there exists a set of t orthogonal Latin squares of order n', then there exists a set of t orthogonal Latin squares of order nn'.*

Proof. We represent the t orthogonal Latin squares of order n and the t orthogonal Latin squares of order n' by an n^2 by $t + 2$ array A and an n'^2 by $t + 2$ array A', respectively. The arrays A and A' are of the type described in Theorem 1.3 and are based on elements labeled $1, 2, \ldots, n$ and $1, 2, \ldots, n'$, respectively. Denote row i of A by

$$(1.14) \quad (a_{i1}, a_{i2}, \ldots, a_{i,t+2})$$

and denote row j of A' by

$$(1.15) \quad (a'_{j1}, a'_{j2}, \ldots, a'_{j,t+2}).$$

We now combine (1.14) and (1.15) into a double-entry row of $t + 2$ components

$$(1.16) \quad ((a_{i1}, a'_{j1}), (a_{i2}, a'_{j2}), \ldots, (a_{i,t+2}, a'_{j,t+2})).$$

The $(nn')^2$ rows of the form (1.16) form an $(nn')^2$ by $t + 2$ array. The entries of this array are the nn' 2-samples of the form

(1.17)

$(1, 1),\ (1, 2),\ \ldots,\ (1, n'),\ \ldots,\ (n, 1),\ (n, 2),\ \ldots,\ (n, n')$.

Now by the structure of the arrays A and A' it follows that the rows of each $(nn')^2$ by 2 subarray of the double-entry array are the $(nn')^2$ 2-samples of the elements (1.17). Then by Theorem 1.3 the double-entry array of size $(nn')^2$ by $t + 2$ is equivalent to a set of t orthogonal Latin squares of order nn'.

THEOREM 1.5. *Let* $n = p_1^{\alpha_1} p_2^{\alpha_2} \ldots p_N^{\alpha_N}$ *be the prime power decomposition of an arbitrary positive integer* n, *where the* p_i *are distinct primes and the* α_i *are positive integers. Let*

(1.18) $\qquad t = \min(p_i^{\alpha_i} - 1) \qquad (i = 1, 2, \ldots, N)$.

Then for $t \geq 2$ *there exists a set of* t *orthogonal Latin squares of order* n.

Proof. By Theorem 1.2 there exists a set of t orthogonal Latin squares of order $p_i^{\alpha_i}$ for each $i = 1, 2, \ldots, N$. The theorem follows by a repeated application of Theorem 1.4.

The preceding theorem establishes the existence of a pair of orthogonal Latin squares of order $n \not\equiv 2 \pmod 4$. In the following section we investigate orthogonal Latin squares of order $n \equiv 2 \pmod 4$.

2. The Euler conjecture. Euler proposed the following problem of the 36 officers. This problem asks for an arrangement of 36 officers of 6 ranks and from 6 regiments in a square formation of size 6 by 6. Each row and each column of this formation are to contain one and only one officer of each rank and one and only one officer from each regiment. We may label the ranks and the regiments from 1 through 6 and assign to each officer a 2-sample of the integers 1 through 6. The first component of the 2-sample designates the officer's rank and the second his regiment. Euler's problem then reduces to the construction of a pair of orthogonal Latin squares of order 6. Euler conjectured in 1782 that there exists no pair of orthogonal Latin squares of order $n \equiv 2 \pmod 4$. Tarry around 1900 verified by a systematic enumeration the validity of Euler's conjecture for $n = 6$. But only recently the combined efforts of Bose, Shrikhande, and Parker culminated in the following theorem.

THEOREM 2.1. *Let $n \equiv 2 \pmod 4$ and let $n > 6$. Then there exists a pair of orthogonal Latin squares of order n.*

This theorem shows that the opposite of the expected state of affairs holds and illustrates the danger of leaping to general conclusions from limited empirical evidence. We cannot go into the intricacies of the proof of Theorem 2.1 here. However, we give a simple and elegant construction for a special case of Theorem 2.1.

THEOREM 2.2. *Let $n \equiv 10 \pmod{12}$. Then there exists a pair of orthogonal Latin squares of order n.*

Proof. Let m be an integer for which there exists a pair of orthogonal Latin squares of order m. Let $i = 0, 1, \ldots, 2m$ and define the vectors

$$A_i = (i, i, \ldots, i),$$
(2.1) $$B_i = (i + 1, i + 2, \ldots, i + m),$$
$$C_i = (i - 1, i - 2, \ldots, i - m).$$

Each of the vectors in (2.1) has m components, and these components are to be regarded as integers $\pmod{2m + 1}$.

We now form the differences (mod $2m + 1$)

$$D = A_i - B_i = (2m, 2m - 1, \ldots, m + 1),$$
$$D' = B_i - A_i = (1, 2, \ldots, m),$$
(2.2)
$$E = A_i - C_i = (1, 2, \ldots, m),$$
$$E' = C_i - A_i = (2m, 2m - 1, \ldots, m + 1),$$
$$F = B_i - C_i = (2, 4, \ldots, 2m),$$
$$F' = C_i - B_i = (2m - 1, 2m - 3, \ldots, 1).$$

In (2.2) it is clear that the $2m$ components of D and D' are $1, 2, \ldots, 2m$ (mod $2m + 1$), and the same holds for E and E', and F and F'. This is the case for each $i = 0, 1, \ldots, 2m$. From the vectors (2.1) we construct the vectors

(2.3)
$$A = (A_0, A_1, \ldots, A_{2m}),$$
$$B = (B_0, B_1, \ldots, B_{2m}),$$
$$C = (C_0, C_1, \ldots, C_{2m}).$$

Then by (2.2),

(2.4)
$$A - B = (D, D, \ldots, D),$$
$$B - A = (D', D', \ldots, D'),$$
$$A - C = (E, E, \ldots, E),$$
$$C - A = (E', E', \ldots, E'),$$
$$B - C = (F, F, \ldots, F),$$
$$C - B = (F', F', \ldots, F').$$

Each of the vectors in (2.3) and (2.4) has $m(2m + 1)$ components. Let

(2.5) $$X = (x_1, x_2, \ldots, x_m)$$

be a permutation of m elements. From the permutation (2.5) we construct the $m(2m + 1)$-sample

(2.6) $$Y = (X, X, \ldots, X).$$

Now from A, B, C, and Y we construct the array

$$(2.7) \qquad G = \begin{bmatrix} A & B & C & Y \\ B & A & Y & C \\ C & Y & A & B \\ Y & C & B & A \end{bmatrix}$$

of size 4 by $4m(2m + 1)$.

Let G' be a 2 by $4m(2m + 1)$ subarray of G. Then G' contains a subarray

$$(2.8) \qquad \begin{bmatrix} A & Y \\ Y & A \end{bmatrix}, \quad \begin{bmatrix} B & Y \\ Y & B \end{bmatrix}, \quad \text{or} \quad \begin{bmatrix} C & Y \\ Y & C \end{bmatrix}.$$

By the structure of A, B, C, and Y this means that G' contains the columns

$$(2.9) \qquad \binom{i}{x_j} \quad \text{and} \quad \binom{x_j}{i}.$$

In (2.9) $i = 0, 1, \ldots, 2m \pmod{2m + 1}$ and $j = 1, 2, \ldots, m$. Also, G' contains a subarray

$$(2.10) \qquad \begin{bmatrix} A & B \\ B & A \end{bmatrix}, \quad \begin{bmatrix} A & C \\ C & A \end{bmatrix}, \quad \text{or} \quad \begin{bmatrix} B & C \\ C & B \end{bmatrix}.$$

If $i \neq j$, then $i - j \pmod{2m + 1}$ is a component of D or D', and the same holds for E or E', and F or F'. If, say, $i - j \pmod{2m + 1}$ appears as a component of D, then $i - j \pmod{2m + 1}$ appears $2m + 1$ times in $A - B$. The positions of $i - j \pmod{2m + 1}$ in $A - B$ are such that G' contains a column of the form

$$(2.11) \qquad \binom{i}{j}.$$

The same situation holds in all cases, and this means that G' contains (2.11) for all $i \neq j$, $i, j = 0, 1, \ldots, 2m \pmod{2m + 1}$.

Now by the assumptions made at the outset of the proof, the integer m is such that there exists a pair of orthogonal

Latin squares of order m. Let H be a 4 by m^2 array on $x_1, x_2, \ldots, x_m$ and the transpose of the type described in Theorem 1.3. We now form the array

$$(2.12) \qquad Z = \begin{bmatrix} & & 0 & 1 & \cdots & 2m \\ G & H & 0 & 1 & \cdots & 2m \\ & & 0 & 1 & \cdots & 2m \\ & & 0 & 1 & \cdots & 2m \end{bmatrix}.$$

This array has 4 rows and

$$(2.13) \quad 4m(2m+1) + m^2 + 2m + 1 = (3m+1)^2$$

columns. The columns of each 2 by $(3m+1)^2$ subarray of Z are the $(3m+1)^2$ 2-samples of $0, 1, \ldots, 2m$ (mod $2m+1$) and $x_1, x_2, \ldots, x_m$. Hence the transpose of Z is an array of the type described in Theorem 1.3. Thus there exists a pair of orthogonal Latin squares of order $n = 3m+1$. By Theorem 1.5 we may take $m \equiv 3 \pmod{4}$, and this yields an $n \equiv 10 \pmod{12}$. Other choices of m yield values of n previously covered by Theorem 1.5.

The preceding construction yields the following pair of orthogonal Latin squares of order 10:

$$A = \begin{bmatrix} 0 & 6 & 5 & 4 & x_3 & x_2 & x_1 & 1 & 2 & 3 \\ x_1 & 1 & 0 & 6 & 5 & x_3 & x_2 & 2 & 3 & 4 \\ x_2 & x_1 & 2 & 1 & 0 & 6 & x_3 & 3 & 4 & 5 \\ x_3 & x_2 & x_1 & 3 & 2 & 1 & 0 & 4 & 5 & 6 \\ 1 & x_3 & x_2 & x_1 & 4 & 3 & 2 & 5 & 6 & 0 \\ 3 & 2 & x_3 & x_2 & x_1 & 5 & 4 & 6 & 0 & 1 \\ 5 & 4 & 3 & x_3 & x_2 & x_1 & 6 & 0 & 1 & 2 \\ 2 & 3 & 4 & 5 & 6 & 0 & 1 & x_1 & x_2 & x_3 \\ 4 & 5 & 6 & 0 & 1 & 2 & 3 & x_2 & x_3 & x_1 \\ 6 & 0 & 1 & 2 & 3 & 4 & 5 & x_3 & x_1 & x_2 \end{bmatrix},$$

$$B = \begin{bmatrix} 0 & x_1 & x_2 & x_3 & 1 & 3 & 5 & 2 & 4 & 6 \\ 6 & 1 & x_1 & x_2 & x_3 & 2 & 4 & 3 & 5 & 0 \\ 5 & 0 & 2 & x_1 & x_2 & x_3 & 3 & 4 & 6 & 1 \\ 4 & 6 & 1 & 3 & x_1 & x_2 & x_3 & 5 & 0 & 2 \\ x_3 & 5 & 0 & 2 & 4 & x_1 & x_2 & 6 & 1 & 3 \\ x_2 & x_3 & 6 & 1 & 3 & 5 & x_1 & 0 & 2 & 4 \\ x_1 & x_2 & x_3 & 0 & 2 & 4 & 6 & 1 & 3 & 5 \\ 1 & 2 & 3 & 4 & 5 & 6 & 0 & x_1 & x_2 & x_3 \\ 2 & 3 & 4 & 5 & 6 & 0 & 1 & x_3 & x_1 & x_2 \\ 3 & 4 & 5 & 6 & 0 & 1 & 2 & x_2 & x_3 & x_1 \end{bmatrix}.$$

3. Finite projective planes. At this point we begin our study of finite projective planes. At first glance these systems appear to be entirely unrelated to the orthogonal Latin squares discussed in the preceding sections. However, in § 4 we show that the two topics are intimately interconnected. A *projective plane* π is a mathematical system composed of entities called "points" and other entities called "lines." The points and the lines are bound together by an "incidence relation." Specifically we assume that there exists a well-defined relation "point P is on line L" or, equivalently, "line L passes through point P," subject to the following postulates.

(3.1) Two distinct points of π are on one and only one line of π.

(3.2) Two distinct lines of π pass through one and only one point of π.

(3.3) There exist four distinct points of π, no three of which are on the same line.

Postulates (3.1) and (3.2) are entirely basic to the system. Postulate (3.3) serves to exclude certain degenerate systems that satisfy only (3.1) and (3.2). The preceding postulates imply the following.

(3.4) There exist four distinct lines of π, no three of which pass through the same point.

It is now clear that every proposition concerning a projective plane has a dual proposition obtained by the appropriate interchange of the words "point" and "line," and the expressions "point P is on line L" and "line L passes through point P." Postulate (3.2) is the dual of (3.1) and (3.4) is the dual of (3.3). This "principle of duality" is of fundamental importance in the theory of projective planes.

THEOREM 3.1. *Let P and P' be two distinct points and let L and L' be two distinct lines of a projective plane π. There exist 1-1 mappings of the points on L onto the points on L', the lines through P onto the lines through P', and the points on L onto the lines through P.*

Proof. Let the notation PQ designate the unique line that passes through two distinct points P and Q of π. Let L and L' be two distinct lines of π. There exists a point O of π not on L or L'. For if all points of π are on L and L', then there are points A and B on L and points C and D on L' such that A, B, C, D satisfy the requirements of (3.3). But then this implies that lines AC and BD pass through a point not on L or L'. Now from a point O not on L or L' we may establish a 1-1 mapping of the points on L onto the points on L'. For if P is a point on L, then the unique line OP passes through a unique point P' on L'. This mapping is 1-1 from all points on L onto all points on L'. The second assertion of the theorem is the dual of the first. Let O be a point of π not on L. Then there is a 1-1 mapping of the points on L onto the lines through O. This is also valid if the point is on L because of the second assertion of the theorem. Note that we have shown that at least three distinct points are on each line of π and at least three distinct lines pass through each point of π.

A projective plane π is called *finite* provided that it contains only a finite number of points. Finite projective planes are of fundamental importance in combinatorial mathematics, and from now on they will play a leading role in our development of the subject. Let L be a line of the finite projective plane π and let the total number of points on L be $n + 1$. The positive integer n is called the *order* of π and is a basic invariant of π.

THEOREM 3.2. *Let π be a finite projective plane of order n. Then the total number of points on an arbitrary line of π as well as the total number of lines through an arbitrary point of π are each equal to $n + 1$. Moreover, π has a totality of $n^2 + n + 1$ points and $n^2 + n + 1$ lines.*

Proof. The first statement of the theorem is immediate from Theorem 3.1 and the definition of order. Let O be a point of π. Then there are exactly $n + 1$ lines through O and there are exactly n points on each of these lines in addition to O. Hence π has a totality of $1 + n(n + 1) = n^2 + n + 1$ points. The dual statement asserts that π has a totality of $n^2 + n + 1$ lines.

Various equivalent sets of postulates are available for a finite projective plane of order n. For example, let π be a mathematical system satisfying (3.2) and (3.3). Let π have a totality of $n^2 + n + 1$ points, with exactly $n + 1$ points on each line and exactly $n + 1$ lines through each point. Then π is a finite projective plane of order n. For if O is a point of π, then there are exactly $n + 1$ lines through O and there are exactly n points on each of these lines in addition to O. This accounts for the totality of $n^2 + n + 1$ points of π. Hence two distinct points of π are on a line of π. This proves (3.1) and π is a finite projective plane of order n.

Finally we note that in many investigations it is convenient to regard the lines of a finite projective plane of order n as certain $(n + 1)$-subsets of the points. The "smallest" projective plane has order $n = 2$. The following 3-subsets of the elements 1, 2, ..., 7 exhibit the 7 lines of the projective plane of order 2:

$$L_1 = \{1, 2, 4\},$$
$$L_2 = \{2, 3, 5\}, \quad L_3 = \{3, 4, 6\}, \quad L_4 = \{4, 5, 7\},$$
$$L_5 = \{5, 6, 1\}, \quad L_6 = \{6, 7, 2\}, \quad L_7 = \{7, 1, 3\}.$$

4. Projective planes and Latin squares. In this section we establish the interconnection between finite projective planes and complete sets of orthogonal Latin squares.

THEOREM 4.1. *Let $n \geq 3$. We may construct a projective plane of order n if and only if we may construct a complete set of $n - 1$ orthogonal Latin squares of order n.*

Proof. Let the projective plane π of order n be given. Let L be a line of π and let $P_1, P_2, \ldots, P_{n+1}$ be the $n + 1$ points on L. Let $Q_1, Q_2, \ldots, Q_{n^2}$ be the n^2 points of π not on L. Let the n lines through P_j not L be labeled 1, 2, ..., n in an arbitrary manner and let this labeling be performed for each $j = 1, 2, \ldots, n + 1$. In particular let Q_iP_j be labeled a_{ij}. Then

(4.1) $$A = [a_{ij}] \qquad (i = 1, 2, \ldots, n^2;$$
$$j = 1, 2, \ldots, n + 1)$$

is an n^2 by $n + 1$ array on 1, 2, ..., n. The rows of each n^2 by 2 subarray of A are the n^2 2-samples of 1, 2, ..., n. For suppose that $a_{ij} = a_{i'j}$ and $a_{ik} = a_{i'k}$, where $i \neq i'$ and $j \neq k$. Then $Q_iP_j = Q_{i'}P_j$ and $Q_iP_k = Q_{i'}P_k$. But then $Q_iQ_{i'}$ passes through P_j and P_k so that $Q_iQ_{i'} = L$. But this contradicts the fact that Q_i and $Q_{i'}$ are not on L.

Sec. 4 PROJECTIVE PLANES AND LATIN SQUARES 93

Hence (4.1) is an array of the type described in Theorem 1.3 and yields a complete set of $n - 1$ orthogonal Latin squares of order n.

Conversely, let (4.1) be an array of the type described in Theorem 1.3. Let the n^2 rows of (4.1) be the "ordinary" points $Q_1, Q_2, \ldots, Q_{n^2}$ and let $P_1, P_2, \ldots, P_{n+1}$ be $n + 1$ "ideal" points. An "ordinary" line L_{ij} passes through P_j and the ordinary points with i in column j of A. The "ideal" line L passes through the ideal points $P_1, P_2, \ldots, P_{n+1}$. Thus far we have defined a configuration π with a totality of $n^2 + n + 1$ points. Each point is on exactly $n + 1$ lines and each line passes through exactly $n + 1$ points. Let L_{ij} and $L_{i'k}$ be two ordinary lines with $j \neq k$. These lines pass through the unique point with i in column j and i' in column k of A. Let L_{ij} and $L_{i'j}$ be two ordinary lines with $i \neq i'$. These lines pass through the unique point P_j. The ordinary line L_{ij} and the ideal line L also pass through the unique point P_j. This proves (3.2). Four points with (1, 1), (1, 2), (2, 1), (2, 2) in the first two positions satisfy (3.3). By the remarks following the proof of Theorem 3.2 it is clear that π is a finite projective plane of order n.

THEOREM 4.2. *Let $n = p^\alpha$, where p is a prime and α is a positive integer. Then there exists a finite projective plane π of order n.*

Proof. This is a consequence of Theorem 1.2 and Theorem 4.1. The plane of order 2 requires special consideration and has been displayed in § 3.

Let n be a positive integer and let d be the largest square dividing n. We write $n = n'd$ and call n' the *squarefree part* of n. If $d = 1$ then n is called *squarefree*, and if $n' = 1$ then n itself is a square. We are now in a position to state the Bruck-Ryser theorem on the nonexistence of finite projective planes.

THEOREM 4.3. *Let* $n \equiv 1$ *or* 2 (mod 4) *and let the square-free part of* n *contain at least one prime factor* $p \equiv 3$ (mod 4). *Then there does not exist a finite projective plane* π *of order* n.

We prove a generalization of Theorem 4.3 in Chapter 8. Evidently Theorem 4.3 excludes geometries for infinitely many values of n, for example, the values $n = 2p$, where p is a prime and $p \equiv 3$ (mod 4). Of course Theorem 4.2 and Theorem 4.3 leave undecided infinitely many values of n. But up to the present time no values of n other than those covered by these theorems have been exhibited or excluded. This has resulted in the development of two opposing points of view. Some maintain that π of order n exists if and only if $n = p^\alpha$, and others maintain that π of order n exists if and only if n is not specifically excluded by Theorem 4.3. The determination of the precise range of values of n is one of the major unsettled issues in combinatorics today. The first undecided case is $n = 10$. The existence of a plane of order 10 requires the construction of a set of 9 orthogonal Latin squares of order 10. But no one has as yet constructed a set of 3 orthogonal Latin squares of order 10.

References for Chapter 7

Papers by MacNeish [8] and Mann [9] deal with the material in § 1. Fundamental papers on the Euler conjecture include Bose and Shrikhande [2; 3] and Parker [11; 12]. Our account of Theorem 2.2 follows Bose, Shrikhande, and Parker [4]. Tarry [18] treats the case $n = 6$. Theorem 4.1 is by Bose [1] and Stevens [17]; Theorem 4.2 by Veblen and Bussey [19]; Theorem 4.3 by Bruck and Ryser [5].

1. R. C. Bose, On the application of the properties of Galois fields to the problem of construction of hyper-Graeco-Latin squares, *Sankhyā*, **3** (1938), 323–338.

2. R. C. Bose and S. S. Shrikhande, On the falsity of Euler's conjecture about the non-existence of two orthogonal Latin squares of order $4t + 2$, *Proc. Nat. Acad. Sci. U. S. A.*, **45** (1959), 734–737.

3. R. C. Bose and S. S. Shrikhande, On the construction of sets of mutually orthogonal Latin squares and the falsity of a conjecture of Euler, *Trans. Amer. Math. Soc.*, **95** (1960), 191–209.

4. R. C. Bose, S. S. Shrikhande, and E. T. Parker, Further results on the construction of mutually orthogonal Latin squares and the falsity of Euler's conjecture, *Canad. Jour. Math.*, **12** (1960), 189–203.

5. R. H. Bruck and H. J. Ryser, The nonexistence of certain finite projective planes, *Canad. Jour. Math.*, **1** (1949), 88–93.

6. M. Hall, Jr., Projective planes, *Trans. Amer. Math. Soc.*, **54** (1943), 229–277.

7. ———, *Projective Planes and Related Topics*, California Institute of Technology, 1954.

8. H. F. MacNeish, Euler squares, *Ann. Math.*, **23** (1922), 221–227.

9. H. B. Mann, The construction of orthogonal Latin squares, *Ann. Math. Stat.*, **13** (1942), 418–423.

10. ———, On orthogonal Latin squares, *Bull. Amer. Math. Soc.*, **50** (1944), 249–257.

11. E. T. Parker, Construction of some sets of mutually orthogonal Latin squares, *Proc. Amer. Math. Soc.*, **10** (1959), 946–949.

12. ———, Orthogonal Latin squares, *Proc. Nat. Acad. Sci. U. S. A.*, **45** (1959), 859–862.

13. ———, Nonextendibility conditions on mutually orthogonal Latin squares, *Proc. Amer. Math. Soc.*, **13** (1962), 219–221.

14. G. Pickert, *Projective Ebenen*, Berlin, Springer-Verlag, 1955.

15. H. J. Ryser, Geometries and incidence matrices, *Slaught Memorial Papers*, no. 4 (1955), 25–31.

16. L. A. Skornyakov, *Projective Planes*, Amer. Math. Soc., translation no. 99, 1953.

17. W. L. Stevens, The completely orthogonalized Latin square, *Ann. Eugen.*, **9** (1939), 82–93.

18. G. Tarry, Le problème de 36 officieurs, *Compte Rendu de l'Association Française pour l'Avancement de Science Naturel*, **1** (1900), 122–123; **2** (1901), 170–203.

19. O. Veblen and W. H. Bussey, Finite projective geometries, *Trans. Amer. Math. Soc.*, **7** (1906), 241–259.

CHAPTER **8**

COMBINATORIAL DESIGNS

1. The (b, v, r, k, λ)-configuration. In this section we introduce combinatorial configurations that are generalizations of the finite projective planes described in the preceding chapter. Let X be a v-set of elements $x_1, x_2, \ldots, x_v$ and let $X_1, X_2, \ldots, X_b$ be b distinct subsets of X. These subsets are called a *balanced incomplete block design* provided they satisfy the following requirements.

(1.1) Each X_i is a k-subset of X.

(1.2) Each 2-subset of X is a subset of exactly λ of the sets $X_1, X_2, \ldots, X_b$.

(1.3) The integers v, k, and λ satisfy $0 < \lambda$ and $k < v - 1$.

Requirements (1.1) and (1.2) are basic to the definition of balanced incomplete block designs, and requirement (1.3) excludes certain degenerate configurations. Balanced incomplete block designs are of great importance in an area of statistics known as the analysis and design of experiments. There the elements are called *varieties* and the sets are called *blocks*. This nomenclature explains our use of the letters v and b. Let $x \in X$ and let x be in exactly r of the subsets $X_1, X_2, \ldots, X_b$. Now consider the $v - 1$ 2-subsets of X that contain x. Both (1.1) and (1.2) may be used to count the occurrences of these $v - 1$ 2-subsets in $X_1, X_2, \ldots, X_b$. If we equate the counts, we obtain

(1.4) $$r(k - 1) = \lambda(v - 1).$$

This tells us that r is another invariant of the balanced incomplete block design. It counts the number of *replications* of an element in the subsets $X_1, X_2, \ldots, X_b$. Since each of the v elements is replicated exactly r times and since each X_i is a k-subset of X, it follows that

(1.5) $$bk = vr.$$

A balanced incomplete block design involves five basic parameters b, v, r, k, and λ, and we henceforth call a balanced incomplete block design a (b, v, r, k, λ)-*configuration*. These five integers are not independent of one another. They are interconnected by (1.4) and (1.5). But (1.4) and (1.5) are by no means the only necessary conditions for the existence of a (b, v, r, k, λ)-configuration. In fact, the central problem in the study of these configurations is the determination of the precise range of values of b, v, r, k, and λ for which configurations exist. This problem is unsolved, and certain special cases of the problem are of fundamental importance in their own right.

Now let

(1.6) $$A = [a_{ij}] \quad (i = 1, 2, \ldots, b; j = 1, 2, \ldots, v)$$

be the incidence matrix of the (b, v, r, k, λ)-configuration. This means, of course, that A is a $(0, 1)$-matrix of size b by v with $a_{ij} = 1$ if x_j is in X_i and $a_{ij} = 0$ if x_j is not in X_i.

The basic properties of the (b, v, r, k, λ)-configuration imply

(1.7) $$AJ = kJ',$$

(1.8) $$A^T A = (r - \lambda)I + \lambda J.$$

Here A^T denotes the transpose of A. The matrix J is the matrix of 1's of order v, J' is the matrix of 1's of size b by v, and I is the identity matrix of order v. Conversely, let $0 < \lambda$ and $k < v - 1$. Then if A is a $(0, 1)$-matrix of size b by v with distinct rows and if A satisfies (1.7) and (1.8), we are assured of the existence of a (b, v, r, k, λ)-configuration with incidence matrix A.

In a $(0, 1)$-matrix if we replace the 1's by 0's and the 0's by 1's, the resulting matrix is called the *complement* of the original matrix. Let A be the incidence matrix of a (b, v, r, k, λ)-configuration and let A' be the complement of A. Then it is easy to verify that A' satisfies

(1.9) $$A'J = k'J',$$

(1.10) $$A'^T A' = (r' - \lambda')I + \lambda' J,$$

where $r' = b - r$, $k' = v - k$, and $\lambda' = b - 2r + \lambda$. The last of these equations and (1.4) and (1.5) imply $(b - r) \times (v - k - 1) = \lambda'(v - 1)$. Hence it follows that for $0 < \lambda$ and $k < v - 1$ we have $0 < \lambda'$ and $k' < v - 1$. Thus A' defines a (b, v, r', k', λ')-configuration. This configuration is called the *complement* of the (b, v, r, k, λ)-configuration.

Let A_1 and A_2 be the incidence matrices for two (b, v, r, k, λ)-configurations. The two configurations are *isomorphic* provided there exists a permutation matrix P of order b and a permutation matrix Q of order v such that

(1.11) $$A_1 = PA_2 Q.$$

Whenever we attempt to classify the (b, v, r, k, λ)-configurations for a specified set of parameters, we count only the distinct configurations in the sense of isomorphism. This procedure is quite proper because isomorphic configurations are the same except for the labeling of the elements and the subsets.

Incidence matrices give us a powerful technique for the investigation of (b, v, r, k, λ)-configurations. We illustrate this point by using incidence matrices to derive an inequality of Fisher.

THEOREM 1.1. *A (b, v, r, k, λ)-configuration has $b \geqq v$.*

Proof. Let A be the incidence matrix of the (b, v, r, k, λ)-configuration. We know that A is of size b by v and we suppose that $b < v$. Then we adjoin $v - b$ rows of 0's to A, and we obtain a square matrix A^* of order v such that

$$(1.12) \qquad A^{*T}A^* = (r - \lambda)I + \lambda J.$$

We now evaluate $\det(A^{*T}A^*)$ in two distinct ways. The matrix A^* contains a row of 0's so that $\det(A^{*T}A^*) = \det(A^{*T})\det(A^*) = 0$. The matrix $(r - \lambda)I + \lambda J$ is of order v and has r in the main diagonal and λ in all other positions. We take this matrix and subtract column 1 from each of the remaining columns. Then we add rows $2, 3, \ldots, v$ to row 1. The resulting matrix has 0's above the main diagonal. Moreover, it follows that

$$(1.13) \quad \det((r - \lambda)I + \lambda J) = (r + (v-1)\lambda)(r - \lambda)^{v-1}.$$

Hence $\det(A^{*T}A^*) \neq 0$ and this contradicts our previous assertion. Thus we must have $b \geqq v$.

By (1.4) and (1.5) a (b, v, r, k, λ)-configuration with $k = 2$ and $\lambda = 1$ has $b = v(v-1)/2$ and $r = v - 1$. Hence the configuration is the set of all 2-subsets of a v-set. A much more interesting state of affairs occurs for the case $k = 3$ and $\lambda = 1$. Let X be a v-set with $v \geqq 3$. A *Steiner triple system* of *order* v is a set of 3-subsets or *triples* of X such that each 2-subset of X is a subset of exactly one triple. The following are Steiner triple systems of orders 3, 7, and 9.

$$(v = 3)$$

$$\{1, 2, 3\}.$$

$$(v = 7)$$

$$\{1, 2, 4\},$$

$$\{2, 3, 5\}, \quad \{3, 4, 6\}, \quad \{4, 5, 7\},$$

$$\{5, 6, 1\}, \quad \{6, 7, 2\}, \quad \{7, 1, 3\}.$$

$$(v = 9)$$

$$\{1, 2, 3\}, \quad \{4, 5, 6\}, \quad \{7, 8, 9\},$$

$$\{1, 4, 7\}, \quad \{2, 5, 8\}, \quad \{3, 6, 9\},$$

$$\{1, 5, 9\}, \quad \{2, 6, 7\}, \quad \{3, 4, 8\},$$

$$\{1, 6, 8\}, \quad \{2, 4, 9\}, \quad \{3, 5, 7\}.$$

Note that the Steiner triple system of order 7 is the same as the projective plane of order 2 in the preceding chapter. It is clear that the Steiner triple systems of order $v > 3$ are the (b, v, r, k, λ)-configurations with $k = 3$ and $\lambda = 1$. Hence by (1.4) and (1.5),

$$(1.14) \qquad b = \frac{v(v-1)}{6}, \qquad r = \frac{(v-1)}{2}.$$

The equations (1.14) imply that *a Steiner triple system has order* $v \geq 3$ *and* $v \equiv 1$ *or* $3 \pmod{6}$. Steiner triple systems have been constructed for all of these values of v. Kirkman made such a construction as early as 1847, and many subsequent constructions have been obtained. The systems are unique for $v = 3, 7$, and 9. There are exactly 2 systems of order 13 and exactly 80 systems of order 15. The number of distinct systems is unknown for $v > 15$. We limit our presentation to the following elementary construction.

THEOREM 1.2. *If there exists a Steiner triple system S_1 of order v_1 and if there exists a Steiner triple system S_2 of order v_2, then there exists a Steiner triple system S of order $v_1 v_2$.*

Proof. Let $\{a_i, a_j, a_k\}$ be in S_1 and let $\{b_r, b_s, b_t\}$ be in S_2. We form a $v_1 v_2$-set of elements c_{ij} ($i = 1, 2, \ldots, v_1$; $j = 1, 2, \ldots, v_2$) and let $\{c_{ir}, c_{js}, c_{kt}\}$ be in S provided (1) $r = s = t$ and $\{a_i, a_j, a_k\}$ is in S_1, or (2) $i = j = k$ and $\{b_r, b_s, b_t\}$ is in S_2, or (3) $\{a_i, a_j, a_k\}$ is in S_1 and $\{b_r, b_s, b_t\}$ is in S_2. It is easy to check that these rules make S a Steiner triple system of order $v_1 v_2$. The triples of S with $r = s = t = 1$ are isomorphic to S_1, and the triples of S with $i = j = k = 1$ are isomorphic to S_2.

Let n be a nonnegative integer. A *Kirkman triple system* of *order* $v = 6n + 3$ is a Steiner triple system of order $v = 6n + 3$ with the following additional stipulation. The set of $b = (2n + 1)(3n + 1)$ triples is partitioned into $3n + 1$ components. Each of the components is a $(2n + 1)$-subset of triples, and each of the $v = 6n + 3$ elements of the triple system appears exactly once in each component. The Steiner triple system of order 3 is a degenerate Kirkman triple system with $n = 0$. The displayed Steiner triple system of order 9 is a Kirkman triple system with $n = 1$. The 12 triples are partitioned into 4 components. Each of the components is displayed as a row of 3 triples, and each of the 9 elements appears exactly once in each row.

Kirkman's famous 15 schoolgirls problem may be formulated as follows. A schoolteacher takes her class of 15 girls on a daily walk. The girls are arranged in 5 rows of 3 each, so that each girl has 2 companions. The problem is to arrange the girls so that for 7 consecutive days no girl will walk with one of her companions in a triplet more than once. This problem is equivalent to the construction of a Kirkman triple system with $n = 2$. The following system fulfills all of these requirements.

$\{1, 2, 5\}$,	$\{3, 14, 15\}$,	$\{4, 6, 12\}$,	$\{7, 8, 11\}$,	$\{9, 10, 13\}$,
$\{1, 3, 9\}$,	$\{2, 8, 15\}$,	$\{4, 11, 13\}$,	$\{5, 12, 14\}$,	$\{6, 7, 10\}$,
$\{1, 4, 15\}$,	$\{2, 9, 11\}$,	$\{3, 10, 12\}$,	$\{5, 7, 13\}$,	$\{6, 8, 14\}$,
$\{1, 6, 11\}$,	$\{2, 7, 12\}$,	$\{3, 8, 13\}$,	$\{4, 9, 14\}$,	$\{5, 10, 15\}$,
$\{1, 8, 10\}$,	$\{2, 13, 14\}$,	$\{3, 4, 7\}$,	$\{5, 6, 9\}$,	$\{11, 12, 15\}$,
$\{1, 7, 14\}$,	$\{2, 4, 10\}$,	$\{3, 5, 11\}$,	$\{6, 13, 15\}$,	$\{8, 9, 12\}$,
$\{1, 12, 13\}$,	$\{2, 3, 6\}$,	$\{4, 5, 8\}$,	$\{7, 9, 15\}$,	$\{10, 11, 14\}$.

2. The (v, k, λ)-configuration. Let X be a v-set of elements $x_1, x_2, \ldots, x_v$ and let $X_1, X_2, \ldots, X_v$ be subsets of X. These subsets are called a (v, k, λ)-*configuration* provided they satisfy the following requirements.

(2.1) Each X_i is a k-subset of X.

(2.2) Each $X_i \cap X_j$ for $i \neq j$ is a λ-subset of X.

(2.3) The integers v, k, and λ satisfy $0 < \lambda < k < v - 1$.

Let

(2.4) $$A = [a_{ij}] \qquad (i, j = 1, 2, \ldots, v)$$

be the incidence matrix of the (v, k, λ)-configuration. Then A is a $(0, 1)$-matrix of order v, and (2.1) and (2.2) imply

(2.5) $$AA^T = B = (k - \lambda)I + \lambda J.$$

Here A^T denotes the transpose of A. The matrix J is the matrix of 1's of order v and I is the identity matrix of order v. Conversely, let $0 < \lambda < k < v - 1$. Then if A is a $(0, 1)$-matrix of order v and if A satisfies (2.5), we are assured of the existence of a (v, k, λ)-configuration with incidence matrix A.

A matrix with real elements is called *normal* provided it commutes with its transpose under multiplication.

THEOREM 2.1. *The incidence matrix A of a (v, k, λ)-configuration is normal. Thus*

$$(2.6) \qquad AA^T = A^TA = B.$$

Proof. Let A be the incidence matrix of the (v, k, λ)-configuration. We know from (1.13) that

$$(2.7) \qquad \det(B) = (k + (v-1)\lambda)(k-\lambda)^{v-1}.$$

Hence $\det(AA^T) = \det(A)\det(A^T) = \det(B) \neq 0$ and $\det(A) \neq 0$. This means that the matrix A is nonsingular and has an inverse denoted by A^{-1}. We know that $AJ = kJ$. Hence it follows that $A^{-1}J = k^{-1}J$. Moreover,

$$(2.8) \qquad AA^TJ = BJ = (k - \lambda + \lambda v)J$$

and

$$(2.9) \qquad A^TJ = (k - \lambda + \lambda v)k^{-1}J.$$

We now take the transpose of both sides of (2.9) and obtain

$$(2.10) \qquad JA = (k - \lambda + \lambda v)k^{-1}J.$$

Hence

$$(2.11) \qquad JAJ = (k - \lambda + \lambda v)k^{-1}vJ.$$

But also

$$(2.12) \qquad JAJ = kvJ,$$

whence it follows that

$$(2.13) \qquad k - \lambda = k^2 - \lambda v.$$

We now substitute (2.13) in (2.10) and obtain

$$(2.14) \qquad JA = kJ.$$

Finally,

$$A^T A = A^{-1}(AA^T)A$$
(2.15)
$$= A^{-1}BA = (k - \lambda)I + \lambda A^{-1}JA$$
$$= (k - \lambda)I + \lambda J = B,$$

and this is the desired conclusion.

THEOREM 2.2. *A (v, k, λ)-configuration is equivalent to a (b, v, r, k, λ)-configuration with $b = v$ and $r = k$.*

Proof. This is an immediate consequence of Theorem 2.1.

In statistics a (v, k, λ)-configuration is called a *symmetrical balanced incomplete block design*. These configurations are encountered in many areas of pure and applied mathematics, and we devote the remainder of this chapter to their study.

THEOREM 2.3. *A finite projective plane of order n is equivalent to a (v, k, λ)-configuration with parameters $v = n^2 + n + 1$, $k = n + 1$, and $\lambda = 1$.*

Proof. This is a consequence of the conclusions in § 3 of Chapter 7 and Theorem 2.1.

In Chapter 7 we established the existence of a projective plane of order $n = p^\alpha$, where p is a prime and α is a positive integer. We also pointed out that thus far finite projective planes have been constructed only for prime power orders. Projective planes are known to be unique for orders $n = 2, 3, 4, 5, 7$, and 8. The number of projective planes of order $n = p^\alpha$ is unknown for $n > 8$.

We now investigate another important class of (v, k, λ)-configurations. A matrix H of order n with entries $+1$ and -1 is called a *Hadamard matrix* provided

(2.16)
$$HH^T = nI,$$

where H^T denotes the transpose of H and I denotes the identity matrix of order n. The matric equation (2.16) implies

(2.17) $$\text{abs. val. det } (H) = n^{n/2}.$$

We mention in passing that Hadamard matrices arise in a very natural way from the following considerations. If a matrix has real elements and if the absolute value of each element does not exceed 1, then a famous inequality of Hadamard asserts that the absolute value of the determinant of the matrix does not exceed $n^{n/2}$. Moreover, one may prove that the value $n^{n/2}$ is attained if and only if the matrix is Hadamard.

The matric equation (2.16) implies that the inverse of H is

(2.18) $$H^{-1} = n^{-1} H^T.$$

Hence it follows that H is normal and

(2.19) $$HH^T = H^T H = nI.$$

If we multiply a row or a column of a Hadamard matrix by -1, the matrix retains the Hadamard property. Hence we may always take the entries in the first row and the first column as $+1$'s. Such a Hadamard matrix is called *normalized*. The normalized Hadamard matrices of orders 1 and 2 are

$$[1],$$

$$\begin{bmatrix} 1 & 1 \\ 1 & -1 \end{bmatrix}.$$

Consider now a normalized Hadamard matrix of order $n \geqq 3$. We may permute the columns so that the second row has the first $n/2$ entries equal to $+1$ and the last $n/2$ entries equal to -1. Let t denote the number of $+1$'s in the first $n/2$ positions of row 3 and let t' denote the number of $+1$'s in the last $n/2$ positions of row 3. Then by (2.16) we have $2t + 2t' = n$ and $2t - 2t' = 0$. Hence $n = 4t$ and we have shown that *a Hadamard matrix of order n has $n = 1, 2$ or $n \equiv 0$ (mod 4)*.

It is conjectured that Hadamard matrices exist for all orders $n \equiv 0$ (mod 4). Impressive techniques are available for the construction of these matrices. For example, for $n \leq 200$ Hadamard matrices have been constructed for all orders $n \equiv 0$ (mod 4) with the exception of $n = 116$ and 188. In what follows we describe a very simple construction. Let $A = [a_{ij}]$ be a matrix of order n and let $A' = [a'_{ij}]$ be a matrix of order n'. Let these matrices have elements in a field F. Then the *direct product* of the matrices A and A' is defined by

$$(2.20) \quad A \times A' = \begin{bmatrix} a_{11}A' & a_{12}A' & \cdots & a_{1n}A' \\ a_{21}A' & a_{22}A' & \cdots & a_{2n}A' \\ \cdot & \cdot & & \cdot \\ \cdot & \cdot & & \cdot \\ \cdot & \cdot & & \cdot \\ a_{n1}A' & a_{n2}A' & \cdots & a_{nn}A' \end{bmatrix}.$$

This is a matrix of order nn'.

THEOREM 2.4. *The direct product of two Hadamard matrices is a Hadamard matrix.*

Proof. Let H be a Hadamard matrix of order n and let H' be a Hadamard matrix of order n'. We may check the row inner products of $H \times H'$ and verify that the requirements for a Hadamard matrix are fulfilled. The following alternative argument utilizes certain valid formal properties of the direct product symbol. Thus

$$(2.21) \quad (H \times H')(H \times H')^T = (H \times H')(H^T \times H'^T)$$
$$= HH^T \times H'H'^T = nI_n \times n'I_{n'} = nn'I_{nn'},$$

where I_n, $I_{n'}$, and $I_{nn'}$ are the identity matrices of orders n, n', and nn', respectively. Note that we have established the existence of a Hadamard matrix of order $n = 2^\alpha$, where α is an arbitrary positive integer.

We now establish the interconnection between Hadamard matrices and (v, k, λ)-configurations.

THEOREM 2.5. *A normalized Hadamard matrix of order $n = 4t \geq 8$ is equivalent to a (v, k, λ)-configuration with parameters $v = 4t - 1$, $k = 2t - 1$, and $\lambda = t - 1$.*

Proof. Let H be a normalized Hadamard matrix of order $n = 4t \geq 8$. We delete row 1 and column 1 of this matrix and we replace the -1's by 0's. This gives us a $(0, 1)$-matrix A of order $v = 4t - 1$. Since H is a normalized Hadamard matrix, it follows that A satisfies the matric equation

$$(2.22) \qquad AA^T = tI + (t-1)J.$$

Hence A is the incidence matrix of a (v, k, λ)-configuration with parameters $v = 4t - 1$, $k = 2t - 1$, and $\lambda = t - 1$. The entire process is reversible, and we may use the incidence matrix A of a (v, k, λ)-configuration with parameters $v = 4t - 1$, $k = 2t - 1$, and $\lambda = t - 1$ to construct the normalized Hadamard matrix of order $n = 4t$.

The complement of a (v, k, λ)-configuration with parameters $v = 4t - 1$, $k = 2t - 1$, and $\lambda = t - 1$ is a (v, k', λ')-configuration with parameters $v = 4t - 1$, $k' = 2t$, and $\lambda' = t$. We call both of these configurations *Hadamard configurations*. It is of interest to note that the (v, k, λ)-configuration with parameters $v = 7$, $k = 3$, and $\lambda = 1$ plays a unique role in our development of the subject. This configuration is simultaneously a Steiner triple system, a finite projective plane, and a Hadamard configuration.

3. A nonexistence theorem. We begin with some preliminary remarks. Let $S = [s_{ij}]$ and $S' = [s'_{ij}]$ be two n by n symmetric matrices with elements in a field F. We say that S and S' are *congruent* over F, written $S \stackrel{c}{=} S'$, provided there exists an n by n nonsingular matrix P with elements in F such that

$$(3.1) \qquad P^T S P = S',$$

where P^T denotes the transpose of P. It is easy to verify that congruence of matrices satisfies the usual requirements of an equals relationship. Thus $S \stackrel{c}{=} S$; $S \stackrel{c}{=} S'$ implies $S' \stackrel{c}{=} S$; $S \stackrel{c}{=} S'$ and $S' \stackrel{c}{=} S^*$ implies $S \stackrel{c}{=} S^*$.

Let $S = [s_{ij}]$ be an n by n symmetric matrix with elements in F and let

$$(3.2) \qquad f = f(x_1, x_2, \ldots, x_n) = \sum_{i,j=1}^{n} x_i s_{ij} x_j$$

be a quadratic form in the indeterminates $x_1, x_2, \ldots, x_n$. We call f the *quadratic form* of the matrix S. Let $S' = [s'_{ij}]$ be an n by n symmetric matrix with elements in F and suppose that $S \stackrel{c}{=} S'$ over F. Then we know that there exists a nonsingular matrix $P = [p_{ij}]$ with elements in F such that $P^T S P = S'$. Now let $y_1, y_2, \ldots, y_n$ be a second set of indeterminates and write

$$(3.3) \qquad x_i = \sum_{j=1}^{n} p_{ij} y_j \qquad (i = 1, 2, \ldots, n).$$

The matrix P is nonsingular so that P has an inverse $P^{-1} = Q = [q_{ij}]$, and (3.3) is equivalent to

$$(3.4) \qquad y_i = \sum_{j=1}^{n} q_{ij} x_j \qquad (i = 1, 2, \ldots, n).$$

Now if we substitute (3.3) into (3.2), we obtain a new quadratic form f' in the indeterminates $y_1, y_2, \ldots, y_n$.

By a direct calculation we may verify that

$$(3.5) \qquad f' = f'(y_1, y_2, \ldots, y_n) = \sum_{i,j=1}^{n} y_i s'_{ij} y_j.$$

In other words, f' is the quadratic form of the matrix S'. The quadratic forms f and f' are called *congruent* over F. We have shown that if $x'_1, x'_2, \ldots, x'_n$ are arbitrary elements in F and if

$$(3.6) \qquad y'_i = \sum_{j=1}^{n} q_{ij} x'_j \qquad (i = 1, 2, \ldots, n),$$

then

$$(3.7) \qquad f(x'_1, x'_2, \ldots, x'_n) = f'(y'_1, y'_2, \ldots, y'_n)$$

is a valid equation in F. Also, if $y_1^*, y_2^*, \ldots, y_n^*$ are arbitrary elements in F and if

$$(3.8) \qquad x_i^* = \sum_{j=1}^{n} p_{ij} y_j^* \qquad (i = 1, 2, \ldots, n),$$

then

$$(3.9) \qquad f(x_1^*, x_2^*, \ldots, x_n^*) = f'(y_1^*, y_2^*, \ldots, y_n^*)$$

is again a valid equation in F.

If A is a matrix of order n and if A' is a matrix of order n' and if these matrices have elements in F, then the *direct sum* of the matrices A and A' is the matrix of order $n + n'$ defined by

$$(3.10) \qquad A \dotplus A' = \begin{bmatrix} A & 0 \\ 0 & A' \end{bmatrix},$$

where the 0's denote zero matrices. If $S_1 \stackrel{c}{=} S'_1$ and if $S_2 \stackrel{c}{=} S'_2$, then it follows without difficulty that

$$(3.11) \qquad S_1 \dotplus S_2 \stackrel{c}{=} S'_1 \dotplus S'_2.$$

The intricacies of the theory of matric congruences depend to a large extent on the nature of the underlying field F. The classical work of Sylvester deals with the problem for the case of the real field, and this is not especially difficult. But the situation is much more complicated for the rational field. This is because the subject is closely related to deep problems in the theory of numbers. We make a few elementary remarks about congruences over the rational field. Let m be a positive integer. Then by Lagrange's four-square theorem we may write

$$(3.12) \qquad m = a_1^2 + a_2^2 + a_3^2 + a_4^2,$$

where a_1, a_2, a_3, a_4 are integers. For our purposes it suffices to know that these four quantities are rational numbers. Now let I_n denote the identity matrix of order n and define

$$(3.13) \qquad H = \begin{bmatrix} a_1 & a_2 & a_3 & a_4 \\ a_2 & -a_1 & a_4 & -a_3 \\ a_3 & -a_4 & -a_1 & a_2 \\ a_4 & a_3 & -a_2 & -a_1 \end{bmatrix}.$$

Then if we multiply H by its transpose, we obtain

$$(3.14) \qquad HH^T = mI_4.$$

Hence we have shown that

$$(3.15) \qquad mI_4 \stackrel{c}{=} I_4$$

over the rational field. By (3.11) this implies

$$(3.16) \qquad mI_n \stackrel{c}{=} I_n$$

for all $n \equiv 0 \pmod 4$.

We return to the (v, k, λ)-configuration. The central problem in the study of these configurations is the determination of the precise range of values of v, k, and λ for which configurations exist. By definition v, k, and λ are integers such that $0 < \lambda < k < v - 1$, and (2.13) asserts

$$(3.17) \qquad k - \lambda = k^2 - \lambda v.$$

Sec. 3   **A NONEXISTENCE THEOREM**

These are necessary conditions on v, k, and λ for the existence of a (v, k, λ)-configuration. Whenever we discuss the existence of these configurations we assume that the parameters satisfy these requirements. The following theorem gives further necessary conditions. No other necessary conditions are known, so that one may conjecture the existence of (v, k, λ)-configurations for all v, k, and λ unless specifically excluded by Theorem 3.1.

THEOREM 3.1. *Let v, k, and λ be integers for which there exists a (v, k, λ)-configuration. If v is even, then $k - \lambda$ equals a square. If v is odd, then the Diophantine equation*

$$(3.18) \qquad x^2 = (k - \lambda)y^2 + (-1)^{(v-1)/2}\lambda z^2$$

has a solution in integers x, y, and z, not all zero.

Proof. Let A be the incidence matrix of the (v, k, λ)-configuration. The matric equation (2.5) asserts

$$(3.19) \qquad AA^T = B = (k - \lambda)I + \lambda J.$$

By (2.7) and (3.17) this implies

$$(3.20) \qquad (\det(A))^2 = \det(B) = k^2(k - \lambda)^{v-1}.$$

Hence $k^2(k - \lambda)^{v-1}$ is a square. Now if v is even, then $k - \lambda$ is a square and this proves our first assertion.

Let v be odd. By (3.19),

$$(3.21) \qquad B \stackrel{c}{=} I$$

over the rational field. We consider first the case $v \equiv 1$ (mod 4). By (3.11) and (3.16),

$$(3.22) \qquad B \stackrel{c}{=} (k - \lambda)I_{v-1} \dotplus I_1.$$

In terms of quadratic forms this means that

$$(3.23) \quad (k - \lambda)(x_1^2 + x_2^2 + \cdots + x_v^2) + \lambda(x_1 + x_2 + \cdots + x_v)^2 = (k - \lambda)(y_1^2 + y_2^2 + \cdots + y_{v-1}^2) + y_v^2.$$

Here

(3.24) $\quad x_i = p_{i1}y_1 + p_{i2}y_2 + \cdots + p_{iv}y_v \quad (i = 1, 2, \ldots, v),$

where the matrix $P = [p_{ij}]$ has rational elements and is nonsingular. In (3.24) if $p_{11} \neq 1$ then we set $x_1 = y_1$, and if $p_{11} = 1$ then we set $x_1 = -y_1$. This induces a dependence relation

(3.25) $\quad\quad\quad y_1 = e_2y_2 + e_3y_3 + \cdots + e_vy_v,$

where $e_2, e_3, \ldots, e_v$ are rational. By (3.24) we have $x_2 = p_2y_2 + p_3y_3 + \cdots + p_vy_v$, where $p_2, p_3, \ldots, p_v$ are rational. Again if $p_2 \neq 1$ then we set $x_2 = y_2$, and if $p_2 = 1$ then we set $x_2 = -y_2$. This induces a dependence relation

(3.26) $\quad\quad\quad y_2 = f_3y_3 + f_4y_4 + \cdots + f_vy_v,$

where $f_3, f_4, \ldots, f_v$ are rational. We continue this process until we reach the stage

(3.27) $\quad\quad\quad y_{v-2} = g_{v-1}y_{v-1} + g_vy_v,$

where g_{v-1} and g_v are rational. By (3.24) we have $x_{v-1} = q_{v-1}y_{v-1} + q_vy_v$, where q_{v-1} and q_v are rational. Finally, we set $x_{v-1} = \pm y_{v-1}$, and this gives us $y_{v-1} = h_vy_v$, where h_v is rational. Thus far we have left y_v unspecified, and we now let y_v equal a nonzero rational. Then $y_{v-1}, y_{v-2}, \ldots, y_1$ are uniquely determined by the various dependence relations and $x_1, x_2, \ldots, x_v$ are uniquely determined by (3.24). Moreover, $x_i^2 = y_i^2$ ($i = 1, 2, \ldots, v - 1$). Then if we substitute these rational numbers into (3.23), we obtain

(3.28) $\quad (k - \lambda)x_v^2 + \lambda(x_1 + x_2 + \cdots + x_v)^2 = y_v^2,$

and this proves the theorem for $v \equiv 1 \pmod{4}$.

Now let $v \equiv 3 \pmod{4}$. The proof of this case requires only minor modifications in the preceding argument. By (3.21), (3.11), and (3.16),

(3.29) $$B \dotplus I_1 \stackrel{c}{=} (k - \lambda)I_{v+1}.$$

In terms of quadratic forms this means that

(3.30) $$(k - \lambda)(x_1^2 + x_2^2 + \cdots + x_v^2) + \lambda(x_1 + x_2 + \cdots + x_v)^2 + x_{v+1}^2 = (k - \lambda)(y_1^2 + y_2^2 + \cdots + y_{v+1}^2).$$

Here

(3.31) $$x_i = p'_{i1}y_1 + p'_{i2}y_2 + \cdots + p'_{i,v+1}y_{v+1}$$
$$(i = 1, 2, \ldots, v + 1),$$

where the matrix $P' = [p'_{ij}]$ has rational elements and is nonsingular. As before we set $x_1 = \pm y_1$ and obtain the dependence relation

(3.32) $$y_1 = e'_2 y_2 + e'_3 y_3 + \cdots + e'_{v+1} y_{v+1},$$

where $e'_2, e'_3, \ldots, e'_{v+1}$ are rational. We set $x_2 = \pm y_2$ and obtain the dependence relation

(3.33) $$y_2 = f'_3 y_3 + f'_4 y_4 + \cdots + f'_{v+1} y_{v+1},$$

where $f'_3, f'_4, \ldots, f'_{v+1}$ are rational. We continue until we reach the stage

(3.34) $$y_{v-1} = g'_v y_v + g'_{v+1} y_{v+1},$$

where g'_v and g'_{v+1} are rational. By (3.31) we have $x_v = q'_v y_v + q'_{v+1} y_{v+1}$, where q'_v and q'_{v+1} are rational. Finally, we set $x_v = \pm y_v$, and this gives us $y_v = h'_{v+1} y_{v+1}$, where h'_{v+1} is rational. We now let y_{v+1} equal a nonzero rational. Then $y_v, y_{v-1}, \ldots, y_1$ and $x_1, x_2, \ldots, x_{v+1}$ are uniquely determined. Moreover, $x_i^2 = y_i^2$ ($i = 1, 2, \ldots, v$). Then if we substitute these rational numbers into (3.30), we obtain

(3.35) $$\lambda(x_1 + x_2 + \cdots + x_v)^2 + x_{v+1}^2 = (k - \lambda)y_{v+1}^2,$$

and this proves the theorem for $v \equiv 3 \pmod{4}$.

Let a and m be nonzero integers and let a and m be relatively prime. The integer a is a *quadratic residue* of m provided the congruence $x^2 \equiv a \pmod{m}$ has a solution and a *quadratic nonresidue* of m provided the congruence $x^2 \equiv a \pmod{m}$ has no solution. Let a, b, and c be nonzero integers. Let these integers be squarefree, relatively prime in pairs, and not all of the same sign. The Diophantine equation

$$(3.36) \qquad ax^2 + by^2 + cz^2 = 0,$$

with a, b, and c fulfilling these requirements, is called a *Legendre equation*. A classical theorem of Legendre asserts that (3.36) has a solution in integers x, y, and z, not all zero, if and only if $-bc$, $-ac$, $-ab$ are quadratic residues of a, b, c, respectively. The necessary conditions of the theorem are obvious. For if (3.36) has a solution in integers x, y, and z, not all zero, then we may assume that these three integers have no prime factor in common. It follows that if a prime p divides a, then p does not divide z. For if $p \mid z$, then $p \mid y$ and $p^2 \mid ax^2$ and $p \mid x$. Thus (3.36) asserts that $(bz^{-1}y)^2 \equiv -bc \pmod{a}$, and the other necessary conditions follow by symmetry. The essential content of Legendre's theorem asserts that these necessary conditions are also sufficient. This portion of the theorem is far from obvious.

It is easy to convert equation (3.18) into a Legendre equation. Let $(k - \lambda)'$ and λ' denote the squarefree parts of $k - \lambda$ and λ, respectively. Now let $d = ((k - \lambda)', \lambda')$, where the notation denotes the positive g. c. d. of $(k - \lambda)'$ and λ'. Then (3.18) has a solution in integers x, y, and z, not all zero, if and only if

$$(3.37) \qquad dx^2 = \frac{(k-\lambda)'}{d} y^2 + (-1)^{(v-1)/2} \frac{\lambda' z^2}{d}$$

has a solution in integers x, y, and z, not all zero. Moreover, equation (3.37) is a Legendre equation. Thus if v is odd and if a (v, k, λ)-configuration exists, then

$$(3.38) \quad (-1)^{(v+1)/2}\frac{\lambda'(k-\lambda)'}{d^2}, \; (-1)^{(v-1)/2}\lambda', \; (k-\lambda)'$$

are quadratic residues of d, $(k-\lambda)'/d$, λ'/d, respectively. Many important (v, k, λ)-configurations have v odd and $(k, \lambda) = 1$. These configurations have $(k - \lambda, \lambda) = 1$ and hence $d = 1$. It is clear that for these configurations the first of the necessary conditions is trivial. The third necessary condition is also trivial. For we always have $k^2 = (k - \lambda) + \lambda v$, and this tells us that $k - \lambda$ is a quadratic residue of λ. But this implies that $(k - \lambda)'$ is a quadratic residue of λ'. Thus the following assertion contains the full content of Theorem 3.1 for (v, k, λ)-configurations with v odd and $(k, \lambda) = 1$. *A (v, k, λ)-configuration with v odd and $(k, \lambda) = 1$ has $(-1)^{(v-1)/2}\lambda'$ a quadratic residue of $(k - \lambda)'$.*

We are now prepared to prove Theorem 4.3 of Chapter 7.

THEOREM 3.2. *Let $n \equiv 1$ or $2 \pmod 4$ and let the square-free part of n contain at least one prime factor $p \equiv 3 \pmod 4$. Then there does not exist a finite projective plane π of order n.*

Proof. A finite projective plane of order n is a (v, k, λ)-configuration with parameters $v = n^2 + n + 1$, $k = n + 1$, and $\lambda = 1$. We have v odd and $(k, \lambda) = 1$. The assumption $n \equiv 1$ or $2 \pmod 4$ means that $(v - 1)/2$ is odd. Hence if the plane exists, then -1 is a quadratic residue of p. But by elementary number theory we know that -1 is a quadratic nonresidue of a prime $p \equiv 3 \pmod 4$.

The existence of Hadamard matrices is conjectured for all orders $n \equiv 0 \pmod 4$. Under these circumstances we would not expect Theorem 3.1 to exclude Hadamard configurations, and this is certainly the case. A Hadamard configuration has parameters $v = 4t - 1$, $k = 2t - 1$, and $\lambda = t - 1$ or $v = 4t - 1$, $k' = 2t$, and $\lambda' = t$. These values lead to the Diophantine equations

(3.39) $\quad x^2 = ty^2 - (t - 1)z^2, \quad x^2 = ty^2 - tz^2.$

The first has the solution $x = y = z = 1$, and the second has the solution $x = 0$ and $y = z = 1$.

4. The matric equation $AA^T = B$. In this section we study the matric equation $AA^T = B$. Throughout the discussion A is a matrix of order v with rational or integral elements and A^T denotes the transpose of A. The matrix B is of order v and is defined by

(4.1) $\quad\quad\quad B = (k - \lambda)I + \lambda J.$

In (4.1) I is the identity matrix of order v and J is the matrix of 1's of order v. We assume that k and λ are integers such that $0 < \lambda < k < v - 1$ and

(4.2) $\quad\quad\quad k - \lambda = k^2 - \lambda v.$

We begin with the following theorem.

THEOREM 4.1. *Let A be a matrix with rational elements such that $AA^T = B$. Then*

(4.3) $\quad\quad A^T A = (k - \lambda)I + \frac{\lambda}{k^2} A^T J A.$

Proof. In § 2 we noted that the matrix B is nonsingular. We now assert that the inverse of B is given by

(4.4) $\quad\quad B^{-1} = \frac{1}{k - \lambda} I - \frac{\lambda}{k^2(k - \lambda)} J.$

For if we multiply (4.4) on the left by B and apply (4.2), we obtain $BB^{-1} = I$. Suppose now that $AA^T = B$. Then $A(A^T B^{-1}) = I$, and since a matrix and its inverse commute we have

(4.5) $\quad\quad\quad A^T B^{-1} A = I.$

Then by (4.4)

(4.6) $$A^T\left(\frac{1}{k-\lambda}I - \frac{\lambda}{k^2(k-\lambda)}J\right)A = I,$$

whence

(4.7) $$A^TA = (k-\lambda)I + \frac{\lambda}{k^2}A^TJA.$$

Theorem 4.1 has a number of interesting consequences. Let s_i denote the sum of column i of A and let t_i denote the sum of the squares of the entries in column i of A. By direct calculation we may verify that

(4.8) $$A^TJA = [s_is_j] \quad (i,j = 1, 2, \ldots, v).$$

Hence Theorem 4.1 implies

(4.9) $$k^2t_i = \lambda s_i^2 + k^2(k-\lambda) \quad (i = 1, 2, \ldots, v).$$

We next observe that if A is a matrix with rational elements such that $AA^T = B$ and if $JA = kJ$, then $A^TA = B$. This assertion is an immediate consequence of Theorem 4.1. An extensive study has been made of the matric congruence $B \stackrel{c}{=} I$ over the rational field. We cannot go into the details of these investigations here. We merely point out that it is known that we always have $B \stackrel{c}{=} I$ over the rational field except for those values of v, k, and λ for which (v, k, λ)-configurations are excluded by Theorem 3.1. Also, it is known that if $B \stackrel{c}{=} I$ over the rational field, there exists a matrix A with rational elements such that $AA^T = A^TA = B$. These investigations are of considerable interest in their own right, but they do not reveal new information on the nonexistence of (v, k, λ)-configurations.

It is natural to study the matric equation $AA^T = B$ for the case in which A has integral elements. We begin with an elementary observation. Let A be a matrix with integral elements such that $AA^T = A^TA = B$. Then we assert that A or $-A$ is the incidence matrix of a (v, k, λ)-configuration. For $A^T A = B$ implies that $t_i = k$, and by (4.9) we have $s_i^2 = k^2$. Now $t_i = k$ and $s_i = k$ imply that the entries in column i of A are 0's and 1's, and $t_i = k$ and $s_i = -k$ imply that the entries in column i of A are 0's and -1's. Each element of B not on the main diagonal of B is equal to the positive integer λ, and this means that columns of both varieties do not occur. Hence A or $-A$ is the incidence matrix of a (v, k, λ)-configuration.

Let S and S' be two n by n symmetric matrices with elements in the ring of integers. The matrix S *integrally represents* S' provided there exists a matrix P of order n with integral elements such that

$$(4.10) \qquad P^T S P = S',$$

where P^T denotes the transpose of P. In particular, the identity matrix I of order n integrally represents S' provided there exists a matrix P of order n with integral elements such that

$$(4.11) \qquad P^T P = S'.$$

It is clear that if a (v, k, λ)-configuration exists, then the identity matrix I of order v integrally represents B. In what follows we prove the converse proposition for certain values of v, k, and λ. Unfortunately the problem of deciding whether or not one matrix integrally represents another is in general a deep, unsettled issue. Further developments here could yield a major advance in our understanding of (v, k, λ)-configurations.

Let A be a matrix with integral elements such that $AA^T = B$. If we multiply a column of A by -1, we do not destroy the matric equation $AA^T = B$. Consequently we may select A so that its column sums are nonnegative. An A fulfilling this requirement is said to be in *normalized form*.

THEOREM 4.2. *Let A be a matrix with integral elements such that $AA^T = B$ and write A in normalized form. If (k, λ) is squarefree and if $k - \lambda$ is odd, then A is the incidence matrix of a (v, k, λ)-configuration.*

Proof. Again let s_i denote the sum of column i of A and let t_i denote the sum of the squares of the entries in column i of A. The matrix A is in normalized form so that each $s_i \geq 0$. Equation (4.9) implies

$$\lambda s_i^2 \equiv 0 \pmod{k^2} \quad (i = 1, 2, \ldots, v). \tag{4.12}$$

Since (k, λ) is squarefree it readily follows from (4.12) and (4.2) that each $s_i \equiv 0 \pmod{k}$. We write $s_i = u_i k$, and (4.9) assumes the form

$$t_i = \lambda u_i^2 + (k - \lambda) \quad (i = 1, 2, \ldots, v). \tag{4.13}$$

Suppose that some $u_i = 0$. Then $s_i = 0$ and by (4.13) we have $t_i = k - \lambda$. But

$$s_i^2 \equiv t_i \equiv k - \lambda \equiv 0 \pmod{2}, \tag{4.14}$$

and this contradicts our assumption that $k - \lambda$ is odd. Hence each $u_i \neq 0$. But $JAA^TJ = JBJ$ implies

$$s_1^2 + s_2^2 + \cdots + s_v^2 = k^2 v, \tag{4.15}$$

whence

$$u_1^2 + u_2^2 + \cdots + u_v^2 = v. \tag{4.16}$$

But if each $u_i \neq 0$, then each $u_i = 1$ and each $s_i = k$. Also, each $t_i = k$. But this implies that A is the incidence matrix of a (v, k, λ)-configuration.

The restriction in Theorem 4.2 that (k, λ) be squarefree is not an especially serious one. Many of the important

(v, k, λ)-configurations satisfy this requirement. For example, finite projective planes and Hadamard configurations with parameters $v = 4t - 1$, $k = 2t - 1$, and $\lambda = t - 1$ both have $(k, \lambda) = 1$. On the other hand, the restriction that $k - \lambda$ be odd excludes from consideration numerous configurations. But we now show that Theorem 4.2 need not be valid for $k - \lambda$ an even integer. In this case (k, λ) squarefree still implies that each $s_i \equiv 0 \pmod{k}$. But we cannot in general conclude that each $u_i \neq 0$.

Let H be a Hadamard matrix of order $n = 2$ or $n \equiv 0 \pmod{4}$. We take H so that its first column is composed only of $+1$'s. We then form the direct sum of H with itself $n + 1$ times. This yields a matrix

$$(4.17) \qquad H' = H \dotplus H \dotplus \cdots \dotplus H$$

of order $n^2 + n$. Now let δ be the row vector of n components with 1 in the first position and 0's elsewhere. We border the matrix H' by an initial row vector

$$(4.18) \qquad \delta' = (\delta, \delta, \ldots, \delta)$$

with $n + 1$ δ's. It is clear that δ' has a totality of $n^2 + n$ components. We then border the resulting array by an initial column vector of $n^2 + n + 1$ components. This vector has 0 in the first position and 1's in all other positions. The resulting matrix A is of order $n^2 + n + 1$, and it is easy to verify that A satisfies the matric equation $AA^T = nI + J$. The matrix A is in normalized form. But A is not the incidence matrix of a projective plane of order n.

We note that a solution of the type just described and an incidence matrix of a projective plane exist simultaneously provided $n = 2^\alpha$. We exhibit these solutions for the case $n = 2$:

THE MATRIC EQUATION $AA^T = B$

(4.19)
$$\begin{bmatrix} 0 & 1 & 0 & 1 & 0 & 1 & 0 \\ 1 & 1 & 1 & 0 & 0 & 0 & 0 \\ 1 & 1 & -1 & 0 & 0 & 0 & 0 \\ 1 & 0 & 0 & 1 & 1 & 0 & 0 \\ 1 & 0 & 0 & 1 & -1 & 0 & 0 \\ 1 & 0 & 0 & 0 & 0 & 1 & 1 \\ 1 & 0 & 0 & 0 & 0 & 1 & -1 \end{bmatrix},$$

(4.20)
$$\begin{bmatrix} 1 & 1 & 0 & 1 & 0 & 0 & 0 \\ 0 & 1 & 1 & 0 & 1 & 0 & 0 \\ 0 & 0 & 1 & 1 & 0 & 1 & 0 \\ 0 & 0 & 0 & 1 & 1 & 0 & 1 \\ 1 & 0 & 0 & 0 & 1 & 1 & 0 \\ 0 & 1 & 0 & 0 & 0 & 1 & 1 \\ 1 & 0 & 1 & 0 & 0 & 0 & 1 \end{bmatrix}.$$

Next we show that integral solutions also exist for $n = 10$. Define

(4.21)

$$H^* = \left[\begin{array}{cc|cccc|cccc} 1 & 1 & 1 & 1 & 1 & 1 & -1 & -1 & -1 & -1 \\ 1 & -1 & 1 & 1 & 1 & 1 & 1 & 1 & 1 & 1 \\ \hline 1 & 1 & & & & & & & & \\ 1 & 1 & & & & & & & & \\ 1 & 1 & & \multicolumn{3}{c}{3I - J} & & \multicolumn{3}{c}{I} & \\ 1 & 1 & & & & & & & & \\ \hline 1 & -1 & & & & & & & & \\ 1 & -1 & & & & & & & & \\ 1 & -1 & & \multicolumn{3}{c}{-I} & & \multicolumn{3}{c}{3I - J} & \\ 1 & -1 & & & & & & & & \end{array} \right],$$

where I denotes the identity matrix of order 4 and J denotes the matrix of 1's of order 4. Then H^* is a matrix of order 10, and it is easy to verify that H^* satisfies $H^*H^{*T} = 10I$, where in this equation I is the identity matrix of order 10. We may now carry out the construction described in the preceding paragraph with H replaced by H^*. This yields a matrix A of order 111 that satisfies the matric equation $AA^T = 10I + J$. But A is far removed as a possible candidate for an incidence matrix of a projective plane of order 10. There appear to exist a wide variety of integral solutions that do not reduce in their normalized form to incidence matrices. In fact, we conjecture that for $v = n^2 + n + 1$, $k = n + 1$, and $\lambda = 1$ such solutions always exist provided only that n is even and $nI + J \overset{c}{=} I$ over the rational field.

We have previously remarked that the existence of Hadamard matrices is conjectured for all orders $n \equiv 0$ (mod 4). We now show that this conjecture could be settled if the theory of integral representations of matrices were more complete. There exists a Hadamard matrix of order 2^α, and the direct product of two Hadamard matrices is a Hadamard matrix. Hence the existence of Hadamard matrices of all orders $4t$ with t odd implies the existence of Hadamard matrices of all orders $n \equiv 0$ (mod 4). But a Hadamard matrix of order $4t \geq 8$ is equivalent to a Hadamard configuration with parameters $v = 4t - 1$, $k = 2t - 1$, and $\lambda = t - 1$. These configurations with t odd are covered by Theorem 4.2. Hence the problem of the existence of Hadamard matrices of all orders $n \equiv 0$ (mod 4) may be regarded as a problem in integral representations.

5. Extremal problems. Thus far our study of (v, k, λ)-configurations has concentrated on the problem of the determination of the precise range of values of v, k, and λ for which these configurations exist. The much more difficult problem of the determination of the actual number of distinct configurations for each choice of v, k, and λ is well beyond the range of present techniques. In this connection we mention that it is conjectured that every projective plane of prime order is unique. There are also important problems involving (v, k, λ)-configurations that are not directly concerned with these two basic problems. Some of these deal with the derivation of deep intrinsic properties of the configurations themselves. For example, let A be the incidence matrix of a projective plane of order n. Do there exist permutation matrices $P \neq I$ and Q such that $PAQ = A$? This is an unsettled question. An affirmative answer would have important consequences in geometry and tell us that every finite projective plane possesses a nontrivial collineation. Many important problems are concerned with the existence and classification of special configurations with additional requirements of one kind or another imposed on them. We mention a particularly interesting instance. An n by n matrix of the form

$$(5.1) \quad A = \begin{bmatrix} a_0 & a_1 & a_2 & \cdots & a_{n-1} \\ a_{n-1} & a_0 & a_1 & \cdots & a_{n-2} \\ a_{n-2} & a_{n-1} & a_0 & \cdots & a_{n-3} \\ \cdot & \cdot & \cdot & & \cdot \\ \cdot & \cdot & \cdot & & \cdot \\ \cdot & \cdot & \cdot & & \cdot \\ a_1 & a_2 & a_3 & \cdots & a_0 \end{bmatrix}$$

is called a *circulant*. The incidence matrix (4.20) of the projective plane of order 2 is a circulant. The incidence matrices of certain (v, k, λ)-configurations can be transformed into circulants by permutations of rows and columns. These configurations are equivalent to the perfect difference sets in number theory and are investigated in

Chapter 9. But first we discuss certain extremal problems that bring together (v, k, λ)-configurations and some of the concepts introduced earlier in this monograph.

Let v and k be fixed integers such that

(5.2) $$1 \leq k \leq v,$$

and let $\mathfrak{A}(K, K)$ denote the class of all $(0, 1)$-matrices of order v with exactly k 1's in each row and column. In Chapter 6 we pointed out that every matrix A in the class $\mathfrak{A}(K, K)$ has per $(A) > 0$. But little is known about the minimal value of per (A) for A in $\mathfrak{A}(K, K)$. If $\mathfrak{A}(K, K)$ contains an incidence matrix of a (v, k, λ)-configuration, then very limited evidence suggests that the permanent of this matrix is small or even minimal in $\mathfrak{A}(K, K)$. Computation becomes prohibitive here even for small values of v. The analogous problem with $\mathfrak{A}(K, K)$ replaced by the doubly stochastic matrices leads to the van der Waerden conjecture of Chapter 5.

Not much is known about the permanent of the incidence matrix of a (v, k, λ)-configuration. However, electronic computers have been used to carry out extensive calculations in this area. Projective planes of orders 2, 3, and 4 have permanents equal to 24, 3852, and 18,534,400, respectively. The permanent of a matrix is invariant under permutations of rows and columns. Consequently it follows that the incidence matrices of two isomorphic (v, k, λ)-configurations have the same permanent. But the incidence matrices of two nonisomorphic (v, k, λ)-configurations on the same parameters v, k, and λ may actually have distinct permanents. Their determinants are of course equal in absolute value.

Let Z denote an incidence matrix of a projective plane of order 2. The matrix Z is of order 7 and has the remarkable property

(5.3) $\quad \text{per }(Z) = \text{abs. val. det }(Z) = 24.$

We now state without proof an interesting theorem that points out the special role played by the matrix Z in the theory of permanents.

THEOREM 5.1. *Let A be a circulant in the class $\mathfrak{A}(K, K)$. If $k > 3$, then*

(5.4) $\quad \text{per }(A) > \text{abs. val. det }(A).$

If $k = 3$ and if

(5.5) $\quad \text{per }(A) = \text{abs. val. det }(A),$

then upon permutations of rows and columns A becomes the direct sum of the matrix Z taken e times. Hence $v = 7e$ and per $(A) = (24)^e$.

Incidence matrices of (v, k, λ)-configurations also arise in extremal problems involving determinants.

THEOREM 5.2. *Let Q be a $(0, 1)$-matrix of order v and let Q contain exactly τ 1's. Define k and λ by*

(5.6) $\quad \tau = kv,$

(5.7) $\quad \lambda = \dfrac{k(k-1)}{v-1},$

and suppose that $0 < \lambda < k < v - 1$. Then

(5.8) $\quad \text{abs. val. det }(Q) \leq k(k - \lambda)^{(v-1)/2},$

and equality holds if and only if Q is the incidence matrix of a (v, k, λ)-configuration.

We omit the proof of Theorem 5.2. Actually this theorem may be generalized in several directions. It seems unlikely that inequalities of the form (5.8) will settle some of the deep arithmetical problems associated with (v, k, λ)-configurations. However, such inequalities are of interest in their own right. Note that Theorem 5.2 implies that if the class $\mathfrak{A}(K, K)$ contains an incidence matrix of a (v, k, λ)-configuration, then the determinant of this matrix is maximal in absolute value over the matrices of the class. This is rather surprising because our earlier remarks indicated that the permanent of such a matrix could conceivably be minimal over the matrices of the class.

Next we point out an interesting connection between finite projective planes and the 1-widths defined in Chapter 6.

THEOREM 5.3. *Let $\mathfrak{A}(K, K)$ be the class with parameters $v = n^2 + n + 1$, $k = n^2$, and $n \geqq 2$. The matrices in $\mathfrak{A}(K, K)$ that are complements of incidence matrices of projective planes of order n have 1-width $\epsilon(1) = 3$ and all other matrices in $\mathfrak{A}(K, K)$ have 1-width $\epsilon(1) = 2$.*

Proof. The proof is almost immediate. Let A be a matrix in $\mathfrak{A}(K, K)$ and form $A^T A$, where A^T denotes the transpose of A. Let λ' be the minimal value and let λ be the average value of the elements of $A^T A$ that are not on the main diagonal of $A^T A$. Evidently

$$(5.9) \qquad \lambda = \frac{n^2(n^2 - 1)}{n^2 + n} = n(n - 1).$$

Suppose that $\lambda' = \lambda$. Then A is the complement of the incidence matrix of a projective plane of order n. In this case the equations $v = n^2 + n + 1$, $k = n^2$, and $\lambda = n(n - 1)$ imply that every v by 2 submatrix of A has exactly one row of 0's. But this means that A has 1-width $\epsilon(1) = 3$. On the other hand, suppose that $\lambda' < \lambda$. Then the equations $v = n^2 + n + 1$ and $k = n^2$ imply that $\lambda' = \lambda - 1$. But this means that a certain v by 2 submatrix of A has no row of 0's. Hence A has 1-width $\epsilon(1) = 2$.

Let $\bar{\epsilon}(1)$ denote the maximal 1-width of the matrices A in the class $\mathfrak{A}(K, K)$ of Theorem 5.3. Then it follows that $\bar{\epsilon}(1) = 3$ if a projective plane of order n exists and $\bar{\epsilon}(1) = 2$ otherwise. In Chapter 6 we let $\tilde{\epsilon}(\alpha)$ be the minimal α-width and we let $\bar{\epsilon}(\alpha)$ be the maximal α-width of the matrices A in the normalized class $\mathfrak{A}(R, S)$. We pointed out that a very efficient procedure was available for the evaluation of $\tilde{\epsilon}(\alpha)$, but that very little was known about the behavior of $\bar{\epsilon}(\alpha)$. Theorem 5.3 illustrates the great complexity of $\bar{\epsilon}(\alpha)$.

References for Chapter 8

A classical paper on (b, v, r, k, λ)-configurations is by Bose [3]. Our proof of Fisher's inequality follows Bose [4]. Selected references on Steiner triple systems include Fort and Hedlund [12]; Hall [17]; Hanani [21]; Moore [34]; Netto [36]; Reiss [40]; Rouse Ball [42]; Skolem [52]. Theorem 2.1 appears in Ryser [43]. The uniqueness of the plane of order 8 is established in Hall, Swift, and Walker [20]. Hadamard matrices are discussed by Baumert, Golomb, and Hall [2]; Brauer [6]; Dade and Goldberg [10]; Gruner [14]; Paley [38]; Todd [56]; Williamson [57; 58]. Section 3 is based on Chowla and Ryser [8] and Bruck and Ryser [7]. A discussion of Legendre's equation is available in Nagell [35]. Section 4 follows Ryser [44] for the most part. For closely allied material see Albert [1]; Goldhaber [13]; Hall and Ryser [19]; Johnsen [28]. For background material on § 4 consult Jones [29] and Taussky [54]. The computation of permanents is discussed by Nikolai [37]. Theorem 5.1 is by Tinsley [55]. Theorem 5.2 appears in Ryser [45; 46] and generalizations of these papers are discussed by Marcus and Gordon [33].

1. A. A. Albert, Rational normal matrices satisfying the incidence equation, *Proc. Amer. Math. Soc.*, **4** (1953), 554–559.

2. L. Baumert, S. W. Golomb, and M. Hall, Jr., Discovery of an Hadamard matrix of order 92, *Bull. Amer. Math. Soc.*, **68** (1962), 237–238.

3. R. C. Bose, On the construction of balanced incomplete block designs, *Ann. Eugen.*, **9** (1939), 353–399.

4. R. C. Bose, A note on Fisher's inequality for balanced incomplete block designs, *Ann. Math. Stat.*, **20** (1949), 619–620.

5. R. C. Bose and D. M. Mesner, On linear associative algebras corresponding to association schemes of partially balanced designs, *Ann. Math. Stat.*, **30** (1959), 21–38.

6. A. Brauer, On a new class of Hadamard determinants, *Math. Zeit.*, **58** (1953), 219–225.

7. R. H. Bruck and H. J. Ryser, The nonexistence of certain finite projective planes, *Canad. Jour. Math.*, **1** (1949), 88–93.

8. S. Chowla and H. J. Ryser, Combinatorial problems, *Canad. Jour. Math.*, **2** (1950), 93–99.

9. W. S. Connor, On the structure of balanced incomplete block designs, *Ann. Math. Stat.*, **23** (1952), 57–71.

10. E. C. Dade and K. Goldberg, The construction of Hadamard matrices, *Michigan Math. Jour.*, **6** (1959), 247–250.

11. R. A. Fisher and F. Yates, *Statistical Tables for Biological, Agricultural, and Medical Research*, London, Oliver and Boyd, 2nd edition, 1943.

12. M. K. Fort, Jr., and G. A. Hedlund, Minimal coverings of pairs by triples, *Pacific Jour. Math.*, **8** (1958), 709–719.

13. J. K. Goldhaber, Integral p-adic normal matrices satisfying the incidence equation, *Canad. Jour. Math.*, **12** (1960), 126–133.

14. W. Gruner, Einlagerung des regulären n-Simplex in den n-dimensionalen Würfel, *Comment. Math. Helv.*, **12** (1939–1940), 149–152.

15. M. Hall, Jr., *Some Aspects of Analysis and Probability*, New York, Wiley, 1958, 35–104.

16. ———, *The Theory of Groups*, New York, Macmillan, 1959.

17. ———, Automorphisms of Steiner triple systems, *IBM Jour. Research and Dev.*, **4** (1960), 460–472.

18. M. Hall, Jr., and W. S. Connor, An embedding theorem for balanced incomplete block designs, *Canad. Jour. Math.*, **6** (1954), 35–41.

19. M. Hall, Jr., and H. J. Ryser, Normal completions of incidence matrices, *Amer. Jour. Math.*, **76** (1954), 581–589.

20. M. Hall, Jr., J. D. Swift, and R. J. Walker, Uniqueness of the projective plane of order eight, *Math. Tables Aids Comput.*, **10** (1956), 186–194.

21. H. Hanani, A note on Steiner triple systems, *Math. Scand.*, **8** (1960), 154–156.

22. ———, The existence and construction of balanced incomplete block designs, *Ann. Math. Stat.*, **32** (1961), 361–386.

23. A. J. Hoffman, M. Newman, E. G. Straus, and O. Taussky, On the number of absolute points of a correlation, *Pacific Jour. Math.*, 6 (1956), 83–96.

24. A. J. Hoffman and M. Richardson, Block design games, *Canad. Jour. Math.*, 13 (1961), 110–128.

25. D. R. Hughes, Collineations and generalized incidence matrices, *Trans. Amer. Math. Soc.*, 86 (1957), 284–296.

26. ———, Generalized incidence matrices over group algebras, *Illinois Jour. Math.*, 1 (1957), 545–551.

27. J. R. Isbell, A class of simple games, *Duke Math. Jour.*, 25 (1958), 423–439.

28. E. C. Johnsen, Matrix rational completions satisfying generalized incidence equations, and integral solutions to the incidence equation for finite projective plane cases of orders $n \equiv 2 \pmod 4$, doctoral dissertation, Ohio State University, 1961.

29. B. W. Jones, *The Arithmetic Theory of Quadratic Forms*, Carus Math. Monograph, no. 10, New York, Wiley, 1950.

30. E. Kleinfeld, Finite Hjelmslev planes, *Illinois Jour. Math.*, 3 (1959), 403–407.

31. K. N. Majumdar, On some theorems in combinatorics relating to incomplete block designs, *Ann. Math. Stat.*, 24 (1953), 377–389.

32. H. B. Mann, *Analysis and Design of Experiments*, New York, Dover, 1949.

33. M. Marcus and W. R. Gordon, Generalizations of some combinatorial inequalities of H. J. Ryser, *Illinois Jour. Math.*, 7 (1963), 582–592.

34. E. H. Moore, Concerning triple systems, *Math. Ann.*, 43 (1893), 271–285.

35. T. Nagell, *Introduction to Number Theory*, New York, Wiley, 1951.

36. E. Netto, *Lehrbuch der Combinatorik*, Leipzig, Teubner, 2nd edition, 1927, reprinted by Chelsea.

37. P. J. Nikolai, Permanents of incidence matrices, *Math. Comput.*, 14 (1960), 262–266.

38. R. E. A. C. Paley, On orthogonal matrices, *Jour. Math. and Physics*, 12 (1933), 311–320.

39. E. T. Parker, On collineations of symmetric designs, *Proc. Amer. Math. Soc.*, 8 (1957), 350–351.

40. M. Reiss, Über eine Steinersche combinatorische Aufgabe welche im 45sten Bande dieses Journals, Seite 181, gestellt worden ist, *Crelle's Jour.*, 56 (1859), 326–344.

41. M. Richardson, On finite projective games, *Proc. Amer. Math. Soc.*, 7 (1956), 458–465.

42. W. W. Rouse Ball, *Mathematical Recreations and Essays* (revised by H. S. M. Coxeter), New York, Macmillan, 1947.

43. H. J. Ryser, A note on a combinatorial problem, *Proc. Amer. Math. Soc.*, 1 (1950), 422–424.

44. ———, Matrices with integer elements in combinatorial investigations, *Amer. Jour. Math.*, 74 (1952), 769–773.

45. ———, Inequalities of compound and induced matrices with applications to combinatorial analysis, *Illinois Jour. Math.*, 2 (1958), 240–253.

46. ———, Compound and induced matrices in combinatorial analysis, *Proc. of Symposia in Applied Math.*, 10 (1960), 149–168.

47. ———, Matrices of zeros and ones, *Bull. Amer. Math. Soc.*, 66 (1960), 442–464.

48. M. P. Schützenberger, A non-existence theorem for an infinite family of symmetrical block designs, *Ann. Eugen.*, 14 (1949), 286–287.

49. S. S. Shrikhande, The impossibility of certain symmetrical balanced incomplete block designs, *Ann. Math. Stat.*, 21 (1950), 106–111.

50. ———, The non-existence of certain affine resolvable balanced incomplete block designs, *Canad. Jour. Math.*, 5 (1953), 413–420.

51. R. Silverman, A metrization for power sets with applications to combinatorial analysis, *Canad. Jour. Math.*, 12 (1960), 158–176.

52. T. Skolem, Some remarks on the triple systems of Steiner, *Math. Scand.*, 6 (1958), 273–280.

53. D. A. Sprott, Note on balanced incomplete block designs, *Canad. Jour. Math.*, 6 (1954), 341–346.

54. O. Taussky, Matrices of rational integers, *Bull. Amer. Math. Soc.*, 66 (1960), 327–345.

55. M. F. Tinsley, Permanents of cyclic matrices, *Pacific Jour. Math.*, 10 (1960), 1067–1082.

56. J. A. Todd, A combinatorial problem, *Jour. Math. Phys.*, 12 (1933), 321–333.

57. J. Williamson, Hadamard's determinant theorem and the sum of four squares, *Duke Math. Jour.*, 11 (1944), 65–81.

58. ———, Note on Hadamard's determinant theorem, *Bull. Amer. Math. Soc.*, 53 (1947), 608–613.

CHAPTER 9

PERFECT DIFFERENCE SETS

1. Perfect difference sets. Let $D = \{d_1, d_2, \ldots, d_k\}$ be a k-set of integers modulo v such that every $a \not\equiv 0 \pmod{v}$ can be expressed in exactly λ ways in the form

(1.1) $$d_i - d_j \equiv a \pmod{v},$$

where d_i and d_j are in D. We suppose further that

(1.2) $$0 < \lambda < k < v - 1.$$

The inequality (1.2) serves only to exclude certain degenerate configurations. A set D fulfilling these requirements is called a *perfect difference set* or, for brevity, a *difference set*. It is easy to verify that

(1.3) $$\lambda = \frac{k(k-1)}{v-1}.$$

This assertion is also an immediate consequence of the following theorem.

THEOREM 1.1. *The perfect difference set D is equivalent to a (v, k, λ)-configuration with incidence matrix a circulant.*

Proof. Let X be the v-set of the integers $0, 1, \ldots, v-1$ modulo v and let D be the given difference set. We define v difference sets

(1.4) $$D_e = \{d_1 + e, d_2 + e, \ldots, d_k + e\}$$
$$(e = 0, 1, \ldots, v-1),$$

where each D_e in (1.4) is a k-subset of X and $D = D_0$. It follows at once from the definition of a difference set that each $D_e \cap D_f$ for $e \neq f$ is a λ-subset of X. Hence the subsets (1.4) are a (v, k, λ)-configuration. Moreover, the incidence matrix for the subsets $D_0, D_1, \ldots, D_{v-1}$ of X is a circulant. The converse proposition also follows without difficulty.

It is clear from the preceding theorem that difference sets may be regarded as special types of (v, k, λ)-configurations, and consequently Theorem 3.1 of Chapter 8 holds for difference sets. However, an arbitrary (v, k, λ)-configuration does not in general yield a difference set. In fact, various values of v, k, and λ are known for which (v, k, λ)-configurations exist and for which difference sets do not exist. A variety of techniques are available for the construction of difference sets. A difference set with $\lambda = 1$ is called *planar*. A planar difference set with $n = k - \lambda$ leads to a projective plane of order n. A projective plane of this type is called *cyclic*. Singer has constructed cyclic projective planes for every order $n = p^\alpha$, where p is a prime and α is a positive integer. It is conjectured that every finite cyclic projective plane must have order $n = p^\alpha$. The correctness of this conjecture has been established for $n \leq 1600$. We mention in passing that it is also conjectured that every finite cyclic projective plane is Desarguesian. Actually this conjecture implies $n = p^\alpha$, but we make no attempt to pursue the interesting topic of Desarguesian planes here. The following table displays planar difference sets for the first few values of n.

n	v	planar difference set
2	7	$\{0, 1, 3\}$
3	13	$\{0, 1, 3, 9\}$
2^2	21	$\{0, 1, 4, 14, 16\}$
5	31	$\{0, 1, 3, 8, 12, 18\}$
7	57	$\{0, 1, 3, 13, 32, 36, 43, 52\}$
2^3	73	$\{0, 1, 3, 7, 15, 31, 36, 54, 63\}$
3^2	91	$\{0, 1, 3, 9, 27, 49, 56, 61, 77, 81\}$
11	133	$\{0, 1, 3, 12, 20, 34, 38, 81, 88, 94, 104, 109\}$

A survey of difference sets with parameters v, k, and λ has been made for k in the interval

$$(1.5) \qquad 3 \leq k \leq 50.$$

Let v, k, and λ be integral parameters that satisfy (1.3), (1.5), and

$$(1.6) \qquad k < \frac{v}{2}.$$

This last assertion is without loss of generality because the complement of a difference set is a difference set. Exactly 268 choices of v, k, and λ fulfill these requirements. Theorem 3.1 of Chapter 8 excludes (v, k, λ)-configurations in 101 cases. This means of course that difference sets do not exist for these 101 choices. The existence of difference sets is established in 46 and the nonexistence in 121 of the remaining 167 cases. Thus there are no undecided cases among these 268 choices.

An impressive array of theorems deal with the construction of difference sets. We prove the following elementary result.

THEOREM 1.2. *Let p be a prime such that $p \equiv 3 \pmod 4$ and $p \geq 7$. Then the set D of the $k = (p-1)/2$ distinct quadratic residues $d_1, d_2, \ldots, d_k \pmod p$ form a difference set with parameters $v = p = 4t - 1$, $k = 2t - 1$, and $\lambda = t - 1$.*

Proof. Let c be a quadratic residue of p. Let d_i and d_j be in D and let $d_i - d_j \equiv 1 \pmod p$. Then $d_i' \equiv cd_i \pmod p$ and $d_j' \equiv cd_j \pmod p$ are in D and satisfy $d_i' - d_j' \equiv c \pmod p$. On the other hand, let d_i' and d_j' be in D and let $d_i' - d_j' \equiv c \pmod p$. Then $d_i \equiv c^{-1}d_i' \pmod p$ and $d_j \equiv c^{-1}d_j' \pmod p$ are in D and satisfy $d_i - d_j \equiv 1 \pmod p$. Thus the totality of d_i and d_j in D such that $d_i - d_j \equiv 1 \pmod p$ is the same as the totality of d_i' and d_j' in D such that $d_i' - d_j' \equiv c \pmod p$. Moreover, if $p \equiv 3$

(mod 4) and if c ranges over the $(p-1)/2$ distinct quadratic residues modulo p, then $-c$ ranges over the $(p-1)/2$ distinct quadratic nonresidues modulo p. Also, $d_i - d_j \equiv c \pmod{p}$ is equivalent to $d_j - d_i \equiv -c \pmod{p}$. This proves that D is a difference set.

The difference set D of Theorem 1.2 yields a Hadamard configuration. The incidence matrix A of this Hadamard configuration is a circulant. This does not mean that the Hadamard matrix H of order $p + 1 = 4t$ associated with A is a circulant. The bordering principle used to obtain H from A destroys the circulant property for H. We digress briefly and remark that the Hadamard matrix of order 4

$$(1.7) \quad \begin{bmatrix} 1 & -1 & -1 & -1 \\ -1 & 1 & -1 & -1 \\ -1 & -1 & 1 & -1 \\ -1 & -1 & -1 & 1 \end{bmatrix}$$

is a circulant. It is conjectured that a Hadamard matrix of order n cannot be a circulant for $n > 4$. We prove that if H is a Hadamard matrix of order n and if H is a circulant, then n must equal a square. For let J be the matrix of 1's of order n. Since H is a Hadamard matrix of order n we have $HH^T = nI$, where H^T is the transpose of H. Since H is a circulant we have $HJ = JH = eJ$, where e is an integer. Hence

$$(1.8) \quad HH^TJ = e^2J = nJ,$$

and this tells us that $n = e^2$.

2. The multiplier theorem. Let X be the v-set of the integers $0, 1, \ldots, v - 1$ modulo v and let $D = \{d_1, d_2, \ldots, d_k\}$ and $D' = \{d'_1, d'_2, \ldots, d'_k\}$ be difference sets on the same parameters v, k, and λ. Let t be an integer such that $(t, v) = 1$ and let s be an arbitrary integer. Then $E =$

$\{td_1, td_2, \ldots, td_k\}$ and $E' = \{d'_1 + s, d'_2 + s, \ldots, d'_k + s\}$ as k-subsets of X are difference sets. We now seek integers t and s such that E and E' are the same k-subset of X. Suppose that such integers can be determined. Then in classifying difference sets it is natural to define D and D' as the same difference set in the sense of isomorphism. We remark that the integers t and s need not exist. For example, it is not difficult to show that for $v = 31$ the quadratic residues modulo 31

(2.1)

$$D = \{1, 2, 4, 5, 7, 8, 9, 10, 14, 16, 18, 19, 20, 25, 28\}$$

and the difference set

(2.2)

$$D' = \{1, 2, 3, 4, 6, 8, 12, 15, 16, 17, 23, 24, 27, 29, 30\}$$

cannot be transformed into the same difference set by the procedure just described. We make no attempt to investigate the uniqueness of difference sets here. However, the preceding discussion provides the motivation for the idea of a multiplier of a difference set and we now study this concept in some detail.

An integer t is called a *multiplier* of the difference set $D = \{d_1, d_2, \ldots, d_k\}$ provided there exists an integer s such that $E = \{td_1, td_2, \ldots, td_k\}$ and $E' = \{d_1 + s, d_2 + s, \ldots, d_k + s\}$ are the same k-subset of X. A multiplier t must satisfy the congruence $td_i - td_j \equiv 1 \pmod{v}$ for some i and j, and hence every multiplier satisfies

$$(2.3) \qquad (t, v) = 1.$$

In the terminology of the preceding discussion it is clear that a multiplier establishes an isomorphism of a difference set with itself. We always have the trivial multiplier $t = 1$. Moreover, it is easy to verify that the multipliers modulo v form a multiplicative group. This group is called the *multiplier group* of the difference set. Multipliers were introduced into the study of difference sets by M. Hall, Jr., and they provide a powerful technique for the derivation of both existence and nonexistence theorems. Unfortunately the theory that we develop does not have a known analogue for (v, k, λ)-configurations.

We begin our investigation of multipliers with some general remarks on congruences. Let $f(x)$, $g(x)$, and $h(x)$ be polynomials with integral coefficients. We write

$$(2.4) \qquad f(x) \equiv g(x) \pmod{h(x)}$$

provided there exists a polynomial $k(x)$ with integral coefficients such that $f(x) - g(x) = k(x)h(x)$. Now let a and b be arbitrary integers and let m be a positive integer. Let a' and b' be nonnegative integers such that $a \equiv a' \pmod{m}$ and $b \equiv b' \pmod{m}$. Then $a \equiv b \pmod{m}$ if and only if $a' \equiv b' \pmod{m}$. Moreover, the congruence

$$(2.5) \qquad a \equiv b \pmod{m}$$

is equivalent to the polynomial congruence

$$(2.6) \qquad x^{a'} \equiv x^{b'} \pmod{x^m - 1}.$$

We now replace the notation (2.6) by the notation

$$(2.7) \qquad x^a \equiv x^b \pmod{x^m - 1}.$$

A "negative" exponent in (2.7) need cause no confusion because (2.7) stands for (2.6). Thus (2.5) is valid for integers a and b if and only if (2.7) is valid for integers a and b.

Now let $D = \{d_1, d_2, \ldots, d_k\}$ be a difference set with parameters v, k, and λ. We define

$$(2.8) \quad \theta(x) \equiv x^{d_1} + x^{d_2} + \cdots + x^{d_k} \pmod{x^v - 1}.$$

Since D is a difference set it follows that

$$(2.9) \quad \theta(x)\theta(x^{-1}) \equiv k + \lambda x + \lambda x^2 + \cdots + \lambda x^{v-1} \pmod{x^v - 1}.$$

For the right side of (2.9) is a sum of terms of the form $x^{d_i - d_j}$ and yields $x^0 = 1$ exactly k times and each $x, x^2, \ldots, x^{v-1}$ exactly λ times. Let

$$(2.10) \quad T(x) = 1 + x + \cdots + x^{v-1}.$$

Then we may rewrite (2.9) in the form

$$(2.11) \quad \theta(x)\theta(x^{-1}) \equiv k - \lambda + \lambda T(x) \pmod{x^v - 1}.$$

We remark that if ϵ is a vth root of unity and if $\epsilon \neq 1$, it follows that $T(\epsilon) = 0$, and (2.11) implies

$$(2.12) \quad k - \lambda = \theta(\epsilon)\theta(\epsilon^{-1}).$$

This equation is of considerable interest in its own right and tells us that difference sets lead to factorization problems involving the vth roots of unity. We return to the multiplier t and observe that t is a multiplier of the difference set D if and only if

$$(2.13) \quad \theta(x^t) \equiv x^s \theta(x) \pmod{x^v - 1}.$$

For the exponents on the left side of the equation are $td_1, td_2, \ldots, td_k \pmod{v}$, and the exponents on the right side of the equation are $d_1 + s, d_2 + s, \ldots, d_k + s \pmod{v}$. We are now prepared to prove the multiplier theorem.

THEOREM 2.1. *Let D be a difference set with parameters v, k, and λ. Let p be a prime divisor of $k - \lambda$ and suppose that $p \nmid v$ and $p > \lambda$. Then p is a multiplier of the difference set D.*

Proof. Since D is a difference set we have

$$(2.14) \quad \theta(x)\theta(x^{-1}) \equiv k - \lambda + \lambda T(x) \pmod{x^v - 1}.$$

Let $f(x)$ be an arbitrary polynomial with integral coefficients. The definition of $T(x)$ implies

$$(2.15) \qquad f(x)T(x) \equiv f(1)T(x) \pmod{x^v - 1}.$$

By hypothesis $p \mid k - \lambda$ and $p \nmid v$. Also, we know that $k - \lambda = k^2 - \lambda v$. This implies that $p \nmid k$, for otherwise $p \mid \lambda v$ and $p \mid \lambda$, whereas $p > \lambda$. Thus

$$(2.16) \qquad k^{p-1} \equiv 1 \pmod{p}.$$

All the coefficients in the expansion of $\theta(x)^p$ apart from the coefficients of $x^{pd_1}, x^{pd_2}, \ldots, x^{pd_k}$ are divisible by p. We now multiply (2.14) by $\theta(x)^{p-1}$ and apply (2.15) and (2.16). Then we may write the resulting expression in the form

$$(2.17) \quad \theta(x^p)\theta(x^{-1}) \equiv \lambda T(x) + pR(x) \pmod{x^v - 1},$$

where $R(x)$ is a polynomial with integral coefficients. The expression $\theta(x^p)\theta(x^{-1})$ of (2.17) regarded as a polynomial of degree less than v has nonnegative integral coefficients. Now $p > \lambda$ and this means that the $R(x)$ of (2.17) regarded as a polynomial of degree less than v has nonnegative integral coefficients. Next we multiply (2.17) by $\theta(x)$ and apply (2.14) and (2.15). This gives

$$(2.18) \quad (k - \lambda)\theta(x^p) \equiv pR(x)\theta(x) \pmod{x^v - 1}.$$

The expressions $\theta(x^p)$, $R(x)$, and $\theta(x)$ of (2.18) regarded as polynomials of degree less than v have nonnegative integral coefficients. Moreover, the structure of (2.18) tells us that $R(x)$ cannot have more than one nonvanishing term. Hence $R(x) = ax^s$, where a and s are nonnegative integers. In (2.18) we may set $x = 1$, and this implies $k - \lambda = pR(1)$. Hence

$$(2.19) \quad (k - \lambda)\theta(x^p) \equiv (k - \lambda)x^s\theta(x) \pmod{x^v - 1},$$

whence p is a multiplier.

Theorem 2.1 establishes the existence of a nontrivial multiplier for every planar difference set because the requirements $p \nmid v$ and $p > \lambda$ are certainly satisfied in this case. The restriction $p > \lambda$ is used critically in the derivation of Theorem 2.1. However, we conjecture that this restriction is not an essential part of the hypothesis. Furthermore, all known difference sets have $(k - \lambda, v) = 1$, so that one may conjecture that every divisor of $k - \lambda$ is a multiplier of the difference set.

We state without proof the following generalization of the multiplier theorem.

THEOREM 2.2. *Let D be a difference set with parameters v, k, and λ. Let d be a divisor of $k - \lambda$ and suppose that $(d, v) = 1$ and $d > \lambda$. Let t be an integer such that for each prime divisor p of d there is an integer j such that $p^j \equiv t \pmod{v}$. Then t is a multiplier of the difference set D.*

A number of worthwhile theorems may be developed from the multiplier concept. We do not investigate these results here. We conclude with a few simple examples that show how multipliers may be used to establish the existence and nonexistence of certain difference sets.

Examples. (a) There exists a difference set with parameters $v = 37$, $k = 9$, and $\lambda = 2$. Let $D = \{d_1, d_2, \ldots, d_k\}$ be a difference set with arbitrary parameters v, k, and λ. We say that D is *fixed* by a multiplier t provided $D = \{td_1, td_2, \ldots, td_k\}$. Suppose that $(t - 1, v) = 1$. Then we assert that there exists a u such that the difference set $D_u = \{d_1 + u, d_2 + u, \ldots, d_k + u\}$ is fixed by t. For the multiplier t maps the difference set D_u onto the difference set

(2.20) $\quad D_{s+tu} = \{d_1 + s + tu, d_2 + s + tu, \ldots, d_k + s + tu\}.$

Hence the multiplier leaves fixed the difference set with u defined by

(2.21) $\quad\quad\quad (t - 1)u \equiv -s \pmod{v}.$

Consider now a difference set with parameters $v = 37$, $k = 9$, and $\lambda = 2$. We have $p = k - \lambda = 7$ and $7 \nmid 37$ and $7 > 2$. Hence by Theorem 2.1, $p = 7$ is a multiplier. Also $(6, 37) = 1$ so there is a difference set fixed by 7. We may multiply the elements of this difference set by a suitable factor so that one element of the difference set is 1. Then the following powers of 7 (mod 37)

(2.22) $\qquad \{1, 7, 9, 10, 12, 16, 26, 33, 34\}$

must be members of the difference set. This is in fact the desired difference set. The construction has also shown that a difference set with these parameters is unique in the sense of isomorphism.

(b) There exists a difference set with parameters $v = 23$, $k = 11$, and $\lambda = 5$. In this case $k - \lambda = 6$, and the two prime divisors of 6 are both less than 5. Hence we cannot apply Theorem 2.1 directly. However, we have $9 \equiv 2^5$ (mod 23) and $9 \equiv 3^2$ (mod 23). This implies that $d = 6$ and $t = 9$ satisfy the requirements of Theorem 2.2. Hence 9 is a multiplier of the difference set. Also $(8, 23) = 1$ so there is a difference set fixed by 9. We may require 1 to be an element of this difference set. Then the following powers of 9 (mod 23)

(2.23) $\qquad \{1, 2, 3, 4, 6, 8, 9, 12, 13, 16, 18\}$

must be members of the difference set. Once again this is the unique difference set with these parameters. It is easy to verify that both 2 and 3 are multipliers of the difference set.

(c) There does not exist a planar difference set with parameters $v = 111$, $k = 11$, and $\lambda = 1$. This is the case of the cyclic projective plane of order 10. By Theorem 2.1 we know that $p = 2$ is a multiplier. Also $(1, 111) = 1$ so there is a difference set fixed by 2. If we apply the multiplier 2 to this difference set, we obtain

(2.24) $\qquad \theta(x^2) \equiv \theta(x) \pmod{x^{111} - 1}.$

Now let ϵ be a cube root of unity and let $\epsilon \neq 1$. We have $111 = 3 \cdot 37$, so that

(2.25) $\qquad \theta(\epsilon^2) = \theta(\epsilon).$

This implies that $\theta(\epsilon) = \theta(\epsilon^{-1})$ is rational. But then (2.12) asserts that $k - \lambda = 10$ is a square. Hence a cyclic projective plane of order 10 does not exist.

References for Chapter 9

Classical papers on difference sets include Singer [12] and Hall [4]. Our account in § 2 is based on Hall [4] and Hall and Ryser [6]. A proof of Theorem 2.2 is available in Hall [5].

1. R. H. Bruck, Difference sets in a finite group, *Trans. Amer. Math. Soc.*, **78** (1955), 464–481.

2. T. A. Evans and H. B. Mann, On simple difference sets, *Sankhyā*, **11** (1951), 357–364.

3. B. Gordon, W. H. Mills, and L. R. Welch, Some new difference sets, *Canad. Jour. Math.*, **14** (1962), 614–625.

4. M. Hall, Jr., Cyclic projective planes, *Duke Math. Jour.*, **14** (1947), 1079–1090.

5. ———, A survey of difference sets, *Proc. Amer. Math. Soc.*, **7** (1956), 975–986.

6. M. Hall, Jr., and H. J. Ryser, Cyclic incidence matrices, *Canad. Jour. Math.*, **3** (1951), 495–502.

7. A. J. Hoffman, Cyclic affine planes, *Canad. Jour. Math.*, **4** (1952), 295–301.

8. D. R. Hughes, Partial difference sets, *Amer. Jour. Math.*, **78** (1956), 650–674.

9. E. Lehmer, On residue difference sets, *Canad. Jour. Math.*, **5** (1953), 425–432.

10. H. B. Mann, Some theorems on difference sets, *Canad. Jour. Math.*, **4** (1952), 222–226.

11. T. G. Ostrom, Concerning difference sets, *Canad. Jour. Math.*, **5** (1953), 421–424.

12. J. Singer, A theorem in finite projective geometry and some applications to number theory, *Trans. Amer. Math. Soc.*, **43** (1938), 377–385.

13. R. G. Stanton and D. A. Sprott, A family of difference sets, *Canad. Jour. Math.*, **10** (1958), 73–77.

14. R. Turyn and J. Storer, On binary sequences, *Proc. Amer. Math. Soc.*, **12** (1961), 394–399.

15. A. L. Whiteman, A family of difference sets, *Illinois Jour. Math.*, **6** (1962), 107–121.

16. K. Yamamoto, Decomposition fields of difference sets, *Pacific Jour. Math.*, **13** (1963), 337–352.

LIST OF NOTATION

$s \in S$	s an element of S, 3
$A \subseteq S$	A a subset of S, 3
$A \subset S$	A a proper subset of S, 4
$P(S)$	Set of all subsets of S, 4
$\emptyset$	Null set, 4
$S \cap T$	Intersection of S and T, 4
$S \cup T$	Union of S and T, 4
n-set	Finite set of $n > 0$ elements, 4
$S \times T$	Product set of S and T, 5
$(a_1, a_2, \ldots, a_r)$	Ordered r-tuple, called sample of size r or r-sample, an r-permutation of n elements if components distinct and selected from an n-set, 5
$P(n, r)$	Number of r-permutations of n elements, 6
$n!$	n-factorial, 6
1-1 mapping	One to one mapping, 6
$G(S)$	Set of all 1-1 mappings of S onto itself, 6
S_n	Symmetric group of degree n, 6
$\{a_1, a_2, \ldots, a_r\}$	Unordered collection of r not necessarily distinct elements, called unordered selection of size r or r-selection, an r-subset if components distinct, and an r-combination of n elements if components distinct and selected from an n-set, 7
$C(n, r) = \binom{n}{r}$	Binomial coefficient, 8
$w(a)$	Weight of a, 17
$[x]$	Greatest integer $\leq x$, 19
g. c. d.	Greatest common divisor, 19
(a, b)	Positive g. c. d. of a and b, 19
$a \mid b$	a divides b, 20
$a \nmid b$	a does not divide b, 20
$\varphi(n)$	Euler φ-function, 20
$\mu(n)$	Möbius function, 21
$\pi(x)$	Number of primes $\leq x$, 22

LIST OF NOTATION

D_n	Number of derangements of n elements, 23
$A = [a_{ij}]$	Rectangular array, called a matrix if entries selected from a field, 24, 25
A^T	Transpose of A, 25
per (A)	Permanent of A, 25
det (A)	Determinant of A, 26
$(0, 1)$-matrix	Matrix with entries the integers 0 and 1, 27
I	Identity matrix, 27
J	Matrix of 1's, 27
U_n	Ménage number, 31
C	$(0, 1)$-matrix with 1's in positions $(1, 2), (2, 3), (3, 4), \ldots, (n, 1)$ and 0's elsewhere, 32
l_n	Number of Latin squares of order n with first row and column in standard order, 36
$P_r(S)$	Set of all r-subsets of S, 38
$N(q_1, q_2, \ldots, q_t, r)$	Minimal positive integer in Ramsey's theorem, 39
N_m	Minimal positive integer in theorem on convex m-gon, 43
SDR	System of distinct representatives, 47
SCR	System of common representatives, 50
ρ	Term rank, 55
$R = (r_1, r_2, \ldots, r_m)$	Row sum vector, 61
$S = (s_1, s_2, \ldots, s_n)$	Column sum vector, 61
τ	Total number of 1's in $(0, 1)$-matrix, 61
$\mathfrak{A} = \mathfrak{A}(R, S)$	Class of all $(0, 1)$-matrices with row sum vector R and column sum vector S, 61
$\bar{A}$	Maximal matrix, 62
$S \prec S^*$	S majorized by S^*, 62
$\tilde{A}$	Special matrix constructed in $\mathfrak{A}$, 63
$\tilde{\rho}$	Minimal term rank of matrices in normalized $\mathfrak{A}$, 70
$\bar{\rho}$	Maximal term rank of matrices in normalized $\mathfrak{A}$, 70
$N_0(Q)$	Number of 0's in $(0, 1)$-matrix Q, 72
$N_1(Q)$	Number of 1's in $(0, 1)$-matrix Q, 72
$\tilde{\sigma}$	Minimal trace of matrices in normalized $\mathfrak{A}$, 76
$\bar{\sigma}$	Maximal trace of matrices in normalized $\mathfrak{A}$, 76
$\epsilon(\alpha)$	α-width, 77
$\tilde{\epsilon}(\alpha)$	Minimal α-width of matrices in normalized $\mathfrak{A}$, 77
$\bar{\epsilon}(\alpha)$	Maximal α-width of matrices in normalized $\mathfrak{A}$, 77
$\mathfrak{A}(K, K)$	Class with $R = S = K = (k, k, \ldots, k)$, 77

LIST OF NOTATION

$GF(p^\alpha)$	Galois field, 80
π	Projective plane, 89
B	Matrix of order v with k in the main diagonal and λ in all other positions, 102
H	Hadamard matrix, 104
$A \times A'$	Direct product of A and A', 106
$S \cong S'$	S congruent to S', 108
$A \dotplus A'$	Direct sum of A and A', 109
$D = \{d_1, d_2, \ldots, d_k\}$	Perfect difference set, 131
t	Multiplier of difference set, 135
$\theta(x)$	$x^{d_1} + x^{d_2} + \cdots + x^{d_k} \pmod{x^v - 1}$, 136
$T(x)$	$1 + x + \cdots + x^{v-1}$, 137

INDEX

Albert, A. A., 127
α-width, 77
   and finite planes, 126, 127
   maximal, 77, 127
   minimal, 77
   of matrix $\tilde{A}$, 77
Array, *see* rectangular array
Asymptotic formula, 36

Bachet, 1
Balanced incomplete block design, *see* (b, v, r, k, λ)-configuration
Baumert, L., 127
Berge, C., 59
Bernoulli, 19
Binomial coefficient, 8
Binomial theorem, 14
Block design, *see* (b, v, r, k, λ)-configuration
Blocks, 96
Bose, R. C., 85, 94, 95, 127, 128
Brauer, A., 127, 128
Bruck, R. H., 93, 94, 95, 127, 128, 141
Bruck-Ryser theorem, 93, 94, 115
Bussey, W. H., 94, 95
(b, v, r, k, λ)-configuration, 96 ff.
   complement of, 98
   Fisher inequality for, 99
   incidence matrix of, 97
   isomorphic, 98

(b, v, r, k, λ)-configuration, necessary conditions for, 97

Chowla, S., 127, 128
Circulant, 123
   and difference sets, 131
   conjecture on Hadamard, 134
   in class $\mathfrak{A}(K, K)$, 125
Class $\mathfrak{A}(K, K)$, 77
   circulants in, 125
   maximal permanent in, 77
   minimal permanent in, 77, 124
Class $\mathfrak{A}(R, S)$, 61 ff.
   existence theorem for, 63
   interchange theorem for, 68
   normalized, 69
   number of matrices in, 65
Collineation, 123
   conjecture on, 123
Column sum vector, 61
   monotone, 61
Combination, 7
Combinatorial mathematics, 2
Common representatives, *see* system of
Complement
   of (b, v, r, k, λ)-configuration, 98
   of $(0, 1)$-matrix, 98
Complete set, 80
   and finite planes, 92
   existence theorem for, 81
Congruence
   of matrices, 108

Congruence, of polynomials, 136
  of quadratic forms, 109
Conjecture
  of Euler, 84 ff.
  of van der Waerden, 59, 77, 124
  on collineations, 123
  on convex m-gons, 44
  on cyclic planes, 132
  on finite planes, 94, 123
  on Hadamard circulants, 134
  on Hadamard matrices, 106, 115, 122
  on integral representations, 122
  on minimal permanents, 124
  on multipliers, 139
  on (v, k, λ)-configurations, 111
Connor, W. S., 128
Convex m-gon, 43
  conjecture on, 44
Convex polygon, 43
Coset, 51
Coxeter, H. S. M., 130
Cyclic plane, 132
  conjectures on, 132
  of order 10, 140

Dade, E. C., 127, 128
Dance problem, 58
Degenerate submatrix, 72
Degree of symmetric group, 6
Derangement, 22 ff.
Desarguesian plane, 132
Determinant, 26
  Hadamard inequality on, 105
  maximal, 125
Dickson, L. E., 16, 37
Difference set, *see* perfect difference set

Direct product, 106
  of Hadamard matrices, 106
Direct sum, 109
Disjoint sets, 4
Distinct representatives, *see* system of
Dominoes, 3
Doubly stochastic matrix, 58
  fundamental theorem on, 58
  van der Waerden conjecture on, 59, 77, 124
Duality, 90
Dulmage, A. L., 60, 78

e^{-1}, 23
Eddington, 7
Element, 3
Enumeration problems, 3
Eratosthenes, 22
Erdös, P., 36, 37, 46
Essential 1, 71
Euler
  conjecture, 84 ff.
  derangement recurrence, 30
  φ-function, 20
  square, *see* orthogonal Latin squares
  36 officers problem, 1, 84
Evans, T., 78
Evans, T. A., 141
Everett, C. J., 59
Existence problems, 2
Extension of Latin rectangle, 52, 66
Ezra, Rabbi Ben, 1

Feller, W., 16, 28
Fibonacci number, 30
Finite field, 80
Finite projective plane, 91 ff.
  and complete sets, 92
  and 1-widths, 126, 127

Finite projective plane, Bruck-Ryser theorem on, 93, 94, 115
  conjecture on collineations of, 123
  conjecture on existence of, 94
  conjecture on uniqueness of, 123
  conjectures on cyclic, 132
  cyclic, 132
  cyclic of order 10, 140
  Desarguesian, 132
  equivalent postulates for, 91
  existence theorem for, 93
  nonexistence theorem for, 93, 94, 115
  number of, 104
  of order 10, 94
  order of, 91
  smallest, 92
Finite set, 4
Fisher inequality, 99
Fisher, R. A., 128
Fixing of difference set, 139
Ford, L. R., Jr., 59, 78
Fort, M. K., Jr., 127, 128
Four-square theorem, 110
Fulkerson, D. R., 59, 78

Gale, D., 78
Galois field, 80
Generalized rule of product, 5
Generalized rule of sum, 4
Gleason, A. M., 46
Goldberg, K., 127, 128
Goldhaber, J. K., 127, 128
Golomb, S. W., 127
Goodman, A. W., 46
Gordon, B., 141
Gordon, W. R., 127, 129
Graeco-Latin square, *see* orthogonal Latin squares

Greatest common divisor, 19
Greenwood, R. E., 46
Group, 6
  and coset decomposition, 51
  as Latin square, 36
  multiplier, 136
  order of, 7
  symmetric, 6
Gruner, W., 127, 128

Haber, R. M., 78
Hadamard configuration, 107
  and difference sets, 133
Hadamard inequality, 105
Hadamard matrix, 104 ff.
  and integral representations, 120, 122
  conjecture on, 106, 115, 122
  conjecture on circulants, 134
  direct product of, 106
  normalized, 105
Hall, M., Jr., 59, 95, 127, 128, 136, 141
Hall, P., 48, 59
Halmos, P. R., 59
Hanani, H., 127, 128
Hardy, G. H., 28
Hedlund, G. A., 127, 128
Higgins, P. J., 60
Hoffman, A. J., 60, 129, 141
Hughes, D. R., 129, 141

Ideal line, 93
Ideal point, 93
Image, 6
Incidence matrix, 53 ff.
  of (b, v, r, k, λ)-configuration, 97
  of (v, k, λ)-configuration, 102
  permanent of, 54, 124
Incidence relation, 89
Inclusion and exclusion formula, 18

Integral representation, 118 ff.
  and Hadamard matrices, 120, 122
  and (v, k, λ)-configurations, 118 ff.
  conjecture on, 122
  unsolved problem on, 118
Interchange, 67
  minimal number of, 68
  theorem on, 68
Interchange theorem, 68
Intermediate term rank, 70
Intersection of sets, 4
Invariant 1, 69
  theorem on, 69
Isbell, J. R., 129
Isomorphism
  of (b, v, r, k, λ)-configurations, 98
  of difference sets, 135

Johnsen, E. C., 127, 129
Jones, B. W., 127, 129

Kaplansky, I., 33, 36, 37
Kirkman, 100
Kirkman triple system, 101
  order of, 101
Kirkman's schoolgirls problem, 1, 101
Kleinfeld, E., 129
König, D., 59, 60
Kuhn, H. W., 60

Lagrange, 110
Laplace expansion, 26
Latin rectangle, 35
  asymptotic formula for, 36
  extension of, 52, 66
  lower bound for, 52, 53
  normalized, 35
  number of, 36, 37
  square, 36

Latin square, 36
  order of, 36
  orthogonal, 79 ff.
  unsolved problem on, 67
Left coset decomposition, 51
Legendre equation, 114
Legendre theorem, 114
Lehmer, E., 141
Line
  ideal, 93
  of matrix, 55
  of projective plane, 89
  ordinary, 93
Lucas, 32

MacNeish, H. F., 94, 95
Magic square, 1
Main diagonal, 25
Majorization, 62
Majumdar, K. N., 129
Mann, H. B., 59, 60, 94, 95, 129, 141
Mapping, 6
  image under, 6
  into, 6
  one to one, 6
  onto, 6
  product of, 6
Marcus, M., 59, 60, 127, 129
Matrix, 25 ff.
  circulant, 123
  congruence of, 108
  direct product of, 106
  direct sum of, 109
  doubly stochastic, 58
  Hadamard, 104 ff.
  incidence, 53 ff.
  integral representation of, 118 ff.
  maximal, 62
  normal, 103
  permutation, 54

INDEX

Matrix, quadratic form of, 108
  zero-one, 27, 53 ff., 61 ff.
Maximal
  α-width, 77, 127
  determinant, 125
  matrix, 62
  permanent, 77
  term rank, 70 ff.
  trace, 76
Ménage number, 32 ff.
Mendelsohn, N. S., 60, 78
Mesner, D. M., 128
Mills, W. H., 141
Minc, H., 60
Minimal
  α-width, 77
  permanent, 59, 77, 124
  term rank, 70, 75
  trace, 76
Möbius function, 21
Monotone column sum vector, 61
Monotone row sum vector, 61
Montmort, 23
Moore, E. H., 127, 129
Multinomial coefficient, 11
Multiplicative law, 26
Multiplicity, 7
Multiplier, 135 ff.
  conjectures on, 139
  examples of, 139, 140
  group, 136
  theorems, 137, 139

Nagell, T., 28, 127, 129
Netto, E., 16, 127, 129
Newman, M., 59, 60, 129
n-factorial, 6
Nikolai, P. J., 127, 129
Nonexistence theorem
  for finite planes, 93, 94, 115
  for (v, k, λ)-configurations, 111

Normal matrix, 103
Normalized
  class, 69
  form of matrix, 118
  Hadamard matrix, 105
  Latin rectangle, 35
n-set, 4
Null set, 4

One to one mapping, 6
  in projective plane, 90
Order
  of finite plane, 91
  of group, 7
  of Kirkman system, 101
  of Latin square, 36
  of square array, 24
  of Steiner system, 99
Ordered partition, 4
Ordered r-tuple, *see* sample
Ordinary line, 93
Ordinary point, 93
Ore, O., 60
Orthogonal Latin squares, 79 ff.
  and finite planes, 92
  as an array, 82
  complete set, 80
  Euler conjecture on, 84 ff.
  existence theorems for, 81 ff.
  for $n \equiv 10 \pmod{12}$, 85
Orthogonal set, *see* orthogonal Latin squares
Ostrom, T. G., 141

Paley, R. E. A. C., 127, 129
Parker, E. T., 85, 94, 95, 129
Partition of set, 4
  ordered, 4
  $(r_1, r_2, \ldots, r_k)$-partition, 10
  unordered, 4
Pascal triangle, 14

Perfect difference set, 131 ff.
　and circulant, 131
　and Hadamard configuration, 133
　conjectures on, 132, 139
　examples of, 139, 140
　fixing of, 139
　isomorphism of, 135
　multiplier of, 135 ff.
　planar, 132
　survey of, 133
Permanent, 25 ff.
　conjectures on, 59, 77, 124
　formula for, 26
　of circulant, 125
　of incidence matrix, 54, 124
Permutation, 5
Permutation matrix, 54
Pickert, G., 95
Pigeon-hole principle, 38
Planar difference set, 132
　table for, 132
Point, 89
　ideal, 93
　ordinary, 93
Polynomial congruence, 136
Prime, 13
Principal submatrix, 44
Principle
　of duality, 90
　of inclusion and exclusion, 18
Problem
　of dance, 58
　of Montmort, 23
　of rooks, 24
　of schoolgirls, 1, 101
　of 36 officers, 1, 84
　of weights, 1
Problème
　des ménages, 32
　des rencontres, 23
Product set, 5

Projective plane, 89 ff.
　duality in, 90
　finite, 91 ff.
　incidence relation in, 89
　line of, 89
　one to one mapping in, 90
　point of, 89
Proper subset, 4

Quadratic form, 108
　congruence of, 109
　of matrix, 108
Quadratic nonresidue, 114
Quadratic residue, 114

Rado, R., 46, 60
Ramsey, F. P., 38, 46
Ramsey's theorem, 38 ff.
　applications, 43 ff.
r-combination, 7
Rectangular array, 24
　main diagonal of, 25
　position in, 24
　size of, 24
　square, 24
　subarray of, 24
　symmetric, 25
　transpose of, 25
Recurrence, 29
Recurrence inequality, 41
Reiss, M., 127, 129
Relatively prime, 19, 20
Replications, 97
Representative, 47
Richardson, M., 129
Right coset decomposition, 51
Riordan, J., 16, 28, 29, 36, 37
Rooks, 24
Rouse Ball, W. W., 127, 130
Row by column rule, 25
Row sum vector, 61
　monotone, 61

INDEX

r-permutation, 5
$(r_1, r_2, \ldots, r_k)$-partition, 10
r-sample, 5
r-selection, 7
r-subset, 4
Rule of product, 5
Rule of sum, 4
Ryser, H. J., 59, 60, 78, 93, 94, 95, 127, 128, 130, 141

Sade, A., 37
Sample, 5
  r-sample, 5
  size of, 5
Schützenberger, M. P., 130
Secondary diagonal, 71
Selection, *see* unordered selection
Set, 3
  disjoint, 4
  element of, 3
  finite, 4
  intersection of, 4
  n-set, 4
  null, 4
  of n elements, 4
  partition of, 4
  proper subset of, 4
  subset of, 3
  union of, 4
Shrikhande, S. S., 85, 94, 95, 130
Sieve formula, 19
Sieve of Eratosthenes, 22
da Silva, 19
Silverman, R., 130
Singer, J., 132, 141
Size
  of array, 24
  of sample, 5
  of unordered selection, 7
Skolem, T., 46, 127, 130
Skornyakov, L. A., 95
Sprott, D. A., 130, 141

Square array, 24
  order of, 24
Squarefree, 93
Squarefree part, 93
Stanton, R. G., 141
Steiner triple system, 99
  number of, 100, 101
  order of, 99
Stevens, W. L., 94, 95
Storer, J., 141
Straus, E. G., 129
Subarray, 24
Subset, 3
  proper, 4
  r-subset, 4
Swift, J. D., 127, 128
Sylvester, 19, 109
Symmetric array, 25
Symmetric group, 6
  degree of, 6
Symmetrical balanced incomplete block design, *see* (v, k, λ)-configuration
System of common representatives, 50 ff.
  application to cosets, 51
  fundamental theorem on, 50
System of distinct representatives, 47 ff.
  fundamental theorem on, 48
  lower bound for, 48
Szekeres, G., 46

Table
  for l_n, 37
  for $N(q_1, q_2, 2)$, 42
  for planar difference sets, 132
Tarry, G., 85, 94, 95
Taussky, O., 127, 129, 130
Term rank, 55 ff.
  fundamental theorem on, 55
  intermediate, 70

Term rank, maximal, 70 ff.
  minimal, 70, 75
  with $\tilde{\rho} = \bar{\rho}$, 76
Tinsley, M. F., 127, 130
Todd, J. A., 127, 130
Touchard, J., 33, 37
Trace, 55
  maximal, 76
  minimal, 76
Transpose, 25
Triple, 99
Triple system
  of Kirkman, 101
  of Steiner, 99
Turyn, R., 141

Unessential 1, 71
Union of sets, 4
Unordered collection, *see* unordered selection
Unordered partition, 4
Unordered selection, 7
  multiplicity of, 7
  r-selection, 7
  size of, 7

Varieties, 96
Vaughan, H. E., 59
Veblen, O., 94, 95

(v, k, λ)-configuration, 102 ff.
  and integral representations, 118 ff.
  and maximal determinants, 125
  and permanents, 124
  conjectures on, 111, 124
  incidence matrix of, 102
  nonexistence theorem for, 111

van der Waerden conjecture, 59, 77, 124
Walker, R. J., 127, 128
Weight, 17
Welch, L. R., 141
Whaples, G., 59
Whiteman, A. L., 141
Williamson, J., 127, 130
Wright, E. M., 28

Yamamoto, K., 36, 37, 141
Yates, F., 128
Yu, Chinese Emperor, 1

Zero-one matrix, 27, 53 ff., 61 ff.
  as incidence matrix, 53 ff.
  class of, 61 ff.
  complement of, 98